GW01454306

Nomads

of the Strait of Gibraltar

Nomads

of the Strait of Gibraltar

Fernando Barrios Partida

A FIELD GUIDE TO BIRD MIGRATION, THE NATURAL PARKS OF THE STRAIT AND LOS ALCORNOCALES, AND THE ROCK OF GIBRALTAR

To those who,
in search of a decent life,
cross the Strait.

Fernando Barrios Partida

Author of Nomads of the Strait of Gibraltar

He was born in Algeciras, Spain, in 1943. It could be that living permanently in contact with the Strait of Gibraltar and the surrounding area of the Campo de Gibraltar had a special influence on the fascination which bird migration holds for him. In his boyhood and youth he went scuba diving off the coast of what is now the Natural Park of the Strait. Later he swapped his harpoon gun for a pair of binoculars. In the 1970's he took part in some studies of soaring birds in the Strait, under Prof. Francisco Bernis, and that was the catalyst that took him away from the undersea world to devote his time almost exclusively to the study and photography of bird migration.

He is a former president of AEFONA (The Spanish Association of Nature Photographers) and his photographs have appeared in such prestigious publications as the National Geographic España and the BBC Magazine, among others. His photographs have also been published in magazines and books in Greece, Portugal, Germany, Italy, France, Belgium, the USA, etc. and, as an expert on the White-rumped Swift (*Apus caffer*), he is a co-author of books such as *El libro rojo de las aves de España, Atlas de las aves reproductoras de España, and the EBCC Atlas of European Breeding Birds.*

index

Photographs and texts:
Fernando Barrios Partida
londra@ono.com

Translation:
Michael Potts
michaelpotts@telefónica.net

Layout and design:
Ana Carnicer
www.carnicermaraia.es

Printed by:
Grafisur
preimpresion@grafisur.com

ISBN 84-934263-4-2
Depósito legal: CA-230/07

Foreword

Keith L. Bildstein, Ph. D.
Sarkis Acopian Director of Conservation Science. Hawk Mountain Sanctuary, Orwigsburg, Pennsylvania, USA

Few days in my life are as clear to me today as when they first happened many years ago. The day I first set eyes on the Strait of Gibraltar is one of them.

It was April 1995 and I was driving through the Parque Natural de los Alcornocales toward Tarifa late in the afternoon with two colleagues when, several kilometers northwest of Los Barrios, our Rover crested a hill that provided us with our first view of the Rock of Gibraltar, an iconic limestone promontory from our child-hoods that all of us recognized immediately. We were approaching our intended destination: The Strait Gibraltar, one of the world's great wildlife corridors and bottlenecks. As we reached Algeciras and crested yet another hill, a second peak-one that we did not recognize-towered in the distance. A glance at our Michelin map of southern Spain indicated that there were no large mountains ahead of us. Perhaps we had turned onto the wrong road and were heading toward Malaga. We quickly agreed that was not possible as Gibraltar was still to our left. The three of us continued on assuming the map was wrong. Another hill-top view south of Algeciras explained the source of our confusion. The largest water-filled crevasse any of us had ever seen dominated the bi-continental landscape surrounding it. Fifth-grade geography sprang into action. The Mediterranean Sea was to our immediate left, the Atlantic Ocean to our far right, and the mountain that towered in front of us was not in southern Spain after all but, rather, was Jebel Musa in northern Morocco. The Strait of Gibraltar, the enigmatic and breathtaking intercontinental intersection that simultaneously connects and separates Europe and Africa and the Atlantic Ocean and the Mediterranean Sea, lay before us. We pulled our car to the side of the road, turned off the ignition, jumped out, ran across the roadway, took several photographs, placed our hands on our hips, and gazed in wonder at the scene before us. We had reached the Pillars of Hercules, the ancient gateway to Ultima Thule; and like many before us, we were overwhelmed. We smiled. Nowhere else on earth does continental geography fall into place so quickly. I still keep one of the grainy snapshots of the Strait that I took on that day. Until now, I showed it to friends in the hopes of conveying my excitement for the place. It never did. My sub-professional photograph is now a historic artifact, happily made obsolete by the stunning series of photographs that make up the "soul" of this spectacular book.

My first visit to The Strait of Gibraltar changed my life, and I have made it a point to return to the region at least annually since. Until now my field work as a raptor biologist and my growing passion for the region's wildlife have fueled my home-comings. *Nomads of the Strait of Gibraltar* by Fernando Barrios ensures that I will now revisit The Strait better equipped and more enlightened than ever before. Photograph-based guides to the world's great ecosystems are the backbones of nature travel. Those that include an insightful and informative text as well as stunning images, allow uninitiated naturalists, professional and amateur alike, along with casual visitors, to rapidly acquaint themselves with a region's landscapes, seascapes, and skyscapes, as well as its wildlife. On the other hand, the lack of a regional guide can turn a potentially positive experience into a frustrating guessing game of missed and unexplored opportunities. Fortunately the latter is no longer the case for The Strait of Gibraltar.

Nomads of the Strait of Gibraltar displays and explains the allure and mystique of this naturally spectacular--and heretofore inadequately presented--region of south-western-most Europe. Authoritative and comprehensive, as well as beautiful, the book provides an entertaining and informative guide to the region's lands and waters, along with the wildlife that regularly move across and live there. I have little doubt that this work will become the standard volume for those interested in the Strait of Gibraltar, and that it will serve as a frequent model for similarly gifted photographer-naturalists wishing to capture the imagery of other special places in the world.

The value of this obviously heartfelt treatise extends well beyond its stunning photographs and personal true-to-life wildlife accounts. Over the years *Nomads of the Strait* will allow those who visit the region an opportunity to "virtually" revisit The Strait at will, and to share with others their visions of and excitement for this truly special natural place. The book also allows readers to appreciate the region for what it is and has been for millennia: a wildlife corridor of unsurpassed global significance, not only for the millions of birds that seasonally transit it while migrating from European and Oceanic breeding grounds to African and Mediterranean wintering areas, but also for the myriads of cetaceans, fishes, and other sea life that seasonally use the waterway while traveling between the Atlantic Ocean and Mediterranean Sea.

Although the "soul" of *Nomads of the Strait* of Gibraltar clearly lies within his breath-taking photographs, which reach out to the reader with stories of their own, the "heart" of Barrios' book lies squarely within a series of crisply written chapter-length essays on Alcornocales Natural Park, the migratory movements of White Storks, raptors, and other soaring migrants, and the human and natural dangers that confront migrants in the region, together with detailed personal accounts of the area's principal wildlife species and important, must-see natural areas.

Part field guide and travel-log, part photographic masterpiece, and part ecological memoir, *Nomads of the Strait* is more than simply the sum of these components. Exceptionally well researched, particularly with regard to historic accounts, this is a book that paints the region's landscapes and fills them with long lists of residents and migrants, and then goes on to explain the biology involved. Throughout, Barrios captures a sense-of-place and spirit for the region that only someone who has lived and breathed it with passion and enthusiasm from all angles and for a very long time could ever achieve. He also presents The Strait with a welcoming wholeness and grandeur that ensures that its many "wonders," including the long-distance migrations of raptors that first attracted me to the site, will continue to amaze and astound those who visit this remarkable special place for a long, long time.

I plan to revisit and reread *Nomads of the Strait of Gibraltar* frequently. I suspect that after paging through it you, too, will do the same. Enjoy your visits.

Prologue to the Spanish edition

The other immigrants

Juan José Téllez Rubio

Fernando Barrios has been scanning the skies of the Strait for many years. I can just imagine him as a wide-eyed youth, when he used to go scuba-diving, blinking through some old wartime binoculars at that cloud-laden sky with which the *levante* wind makes this vertiginous coast easier on the eye: indeed, the road from Tarifa to Algeciras has been described by Rafael Zapatero, the painter, as nothing less than heroic! The dimensions of the scenery have been determined by the gods, not by humans. This is the land of legend and prodigy, from the saga of Ulysses to the peripeteia of the poet Avieno. Yet, over the last few years, this coastline has also been witnessing a massive exercise in survival: that of clandestine immigration across its waters, of thousands of men, women and children following a route established before history began, when the first "European", who was in fact from Africa and female, came to this part of the world.

Since ancient times, birds have been the most frequent "wetbacks" of this unique crossroads of two seas, from White Storks to Starlings, negotiating headwinds, tailwinds, side winds and rebounding winds in this area of predominating east and west winds, where you can see flocks of Kites trying to cross over and over again or Starlings which often leave squares and trees devastated in their wake. All of us, just like Fernando Barrios, have at some time stopped to observe the arrogant figure of a Honey Buzzard or the gentle flapping of the White Stork. But we are also shocked when we read in the newspapers of birds killed flying into wind turbine blades, or flamingos being shot by so-called "hunters" on the banks of the river Palmones, or protected birds being trapped just for the sake of having a stuffed trophy. In Rota,

Mao-Tse Tung`s spies even used carrier pigeons to transport military secrets, but the romantic travellers of yesteryear, like Francis Carter, noted down as early as 1772 that "vultures that came from Africa in the spring and flew over the Rock without stopping, returned each autumn".

This to-ing and fro-ing of birds is nothing new and they may have acquired this aspect of their life cycle during the glacial changes of the Quaternary period. Furthermore, as can be seen from early rock carvings, they lived in harmony with primitive Man, who stood on two legs, but hunched with the fear he must have felt at spending long nights with only a fire to protect him from the roaming beasts. Nowadays, and according to data from the Migres Programme, which the regional Department of the Environment has set up under the guidance of the Spanish Ornithological Association, "a total of 200 species of birds use the Strait of Gibraltar in their migratory movements, be they transcontinental or transoceanic and in massive or small numbers." Harriers or Cranes, Kestrels or Ospreys, Seagulls or Larks – all of them form a roving colony which is a million strong, according to the annual censuses. Many of them are soaring birds, breeding in Europe and wintering in Africa, in a transient ceremony which has a lot to do with their mating habits. This book doesn't only discuss birds, but also refers to their surroundings: majestic ferns, roe deer waiting for poachers, pleasant wild oak trees accompanying cork trees, unscrupulous woodcutters, foxes and badgers, shrews or lizards, fungi which spring up like mushrooms in the Natural Park of Los Alcornocales whose name refers to the cork tree which, in Fernando Barrios' opinion "is on death row, and is both old and defenceless…" What he is saying is that it is not possible to divide nature up into plots or put it into airtight compartments as if we were not all for one and one for all.

Of course the author lays the main emphasis of his book on the wings that, in his mind's eye, brought and carried away dreams and legends. In this personal and untransferable world, where there is lush vegetation but no wild boar or wolf, he describes in detail the book of the hours of these birds that sometimes pause on their journey, between Valdevaqueros and El Tolmo, the beach at La Atunara, Punta Carnero, Guadalmesí, the Rock, Punta Camarinal, Bolonia, Los Lances beach and Tarifa island – judging by the different itineraries described by those well practised in bird observation. To keep track of them, Fernando Barrios has sometimes gone to the most unusual lengths to capture that special detail of these airborne travellers with his camera. On one occasion he even went as far as dressing up as a White Stork: "I asked my friend Antonio Luque if he could make a mobile hide or White Stork costume and, although somewhat surprised, he readily agreed," Barrios tells us further on. "During July and early August, on the flat roof of his house, Antonio started assembling a structure made up of copper tubing, wire and aluminium. Gradually it began to take shape and come to life and the result was a rather large (very well fed) White Stork, but so heavy that some metal supports had to be fitted for me to rest on after walking for a long time." He was transported, rather like a float in a parade, to the lagoon where the storks stopped off and they saw a large group. "Just to make the disguise complete I put on some red panty hose, the same shade as the storks' legs." As he approached them a few took off and flew away: "Little by little the enormous group disappeared until there was just one left with its back to me but my being there, which had scared off the rest of the storks, didn't seem to trouble it. This cheered me up no end and I thought that if one could tolerate my presence then so could a lot more and it would just be a case of changing strategy. Very slowly I got nearer and nearer to its right side and, from only a few metres away, I took several photos. Suddenly I realized that there was something wrong with its right eye and it couldn't see me. I must have moved clumsily because suddenly it turned towards me and, on seeing me with its good eye, it flew off, startled, with its heart in its mouth."

Barrios doesn't just cultivate his mind, he is also something of an adventurer. But his vocation is definitely the public eye and while others observe the daily reality in the cities, hearts broken by unpleasant events, the madness of idylls and the pomp which comes with power, he prefers to contemplate these silent passengers in time, which over the centuries have carried news of floods or remote epidemics in their beaks. The photographer sees them and so does the man in the street, perhaps wondering what is happening to this place that they and we share; in other words this ancient planet is adrift and global warming is already causing serious problems to their routes and habits. This is perhaps a foretaste of those issues which will occupy and worry us humans as the climate changes and the icecaps melting at the Poles become a reality and not just a simple prediction, according to some people in important places, of a pessimistic group of scientists.

Apart from the above considerations, this book is an interesting pointer towards other works, not necessarily bibliographical – e.g. the splendid TV documentary called "El latido del bosque" (The heartbeat of the forest) directed by Joaquín Gutiérrez Acha – and it is also a magnificent roadmap of undiscovered treasures, into which anyone may venture, from the maze of paths around Punta Carnero to the Garganta del Capitán (Captain`s Gorge).

The constant message in the following pages is this: the contrast between the wonder that nature has given to us and what we are in serious danger of losing forever. Barrios often resorts, in an apparently off-handed way but with heartfelt pain, to telling us what might have been but wasn't or mentions what used to be but is no longer. Perhaps this is the message that the birds flying overhead transmitted to him in his youth when he stopped to watch them before diving into the waters of the Strait.

Introduction

At the end of the fifties and, above all, in the sixties, on those days when I used to spend hours underwater fishing in the Strait, I would see lots of birds of prey flying to Morocco or coming from that direction. I realized they were migrating, but knew nothing about their phenology nor could I tell which was which, because I only watched them with the naked eye. In the early sixties my father gave me some Zeiss binoculars, which he had had since the Civil War and they were one of those objects which, as the saying goes, changed my life. So, when I emerged from the sea after several exhausting but nevertheless gratifying hours of skin-diving (not only because of the fish I caught but also the fantastic time I had watching the behaviour of different species), I would pick up my binoculars to watch the strange boats crossing the Strait, the then almost deserted Moroccan coast and, of course, the birds. I used to imagine these magnificent animals, most of them raptors, arriving in my country to breed and then flying back, together with their young (that's what I thought at the time) to spend the winter in warmer climes.

Once when I was swimming out to sea – that was when we *frog-men*, as we were called, couldn't afford boats to take us fishing – to the isle of Las Palomas, just off the Punta Carnero lighthouse, I saw a beautiful, large, white bird, perched on a little rock jutting a few centimetres out of the water next to the island. I thought it must be a large duck, so, with my head just above water and my body submerged as much as possible, I went towards it to get a better look and see if it was injured. When I was about ten metres away I had a clear view, although the bird stared back at me, obviously confused and not realizing I was a human being. I could see it was an Egyptian vulture, its white plumage, yellowish face and penetrating eyes fascinating me for several long seconds before it rose up, flying just a few metres over my head towards the beach on which, quite unexpectedly, it didn't land. Lots of times I saw raptors which I couldn't identify, until one day my friend Manuel Español mentioned a book (Peterson's Bird Guide) and an association in Madrid whose members studied birds (SEO - The Spanish Ornithological Society, which I later joined). Armed with this book, which I bound in plastic to guard against inclement weather, I started to observe birds more painstakingly, in order to identify them. The more birds I identified, the more interested I became in ornithology, to the detriment of my underwater fishing. And, not satisfied with what I could see on the coast, I started to explore the inland sierras of the Campo de Gibraltar region. It was in 1974, after a few years spending more time in the country than in the sea, when I lay down my harpoon for good and took up the binoculars.

In 1972 I took part in the very first organized campaign in the Strait to study bird migration and, on meeting people who spoke to me of the wonders of the area which could only be seen in one or two places in the world, I began to realize how lucky and privileged I was to live here. I found out that the forests of Los Alcornocales were an exception, not a proliferation, and so I began to think that, in some way, this natural wealth which we who live here enjoy, should be made known to and shared with people who have never heard of it, but who would be so impressed that they would want to visit these forests and experience for themselves the amazing spectacle of bird migration. With the publication of this work, one of my constant desires as a naturalist and photographer has been fulfilled: that of promoting the botanical, zoological and scenic values of one of the most interesting areas of the Palaearctic region, the Campo de Gibraltar.

The aim of this book is to inform and, I hope, educate, those who come down to the coast of the Campo de Gibraltar to watch a unique, though cyclic, spectacle: that of the migration of soaring birds whose uncontrollable desire it is to cross the Strait of Gibraltar. But I don't just want readers to learn how to identify the silhouettes of soaring birds in flight. I would like them to know something about the complicated and unpredictable meteorology of the area, to realize the importance of the two natural parks next to the Strait, to learn where to position themselves in the complicated relief of the area in order to observe the migratory flow, to know which are the most favourable times of year to locate certain species, and to discover unknown aspects of their passage. Then, last but not least, I would like them, in this work, to find the answer to most of their questions.

This book is made up of eight chapters. *Chapter I: The Natural Park of the Strait.* This is devoted to the littoral area, a key one in migration, converged upon by the prenuptial migrants, most of them tired after a long, exhausting journey. Very often large birds such as Griffon vultures or Short-toed Eagles have to rest before continuing to their final destination. It is also a springboard at the edge of the Strait, where they have to struggle against strong winds, many of them for the first time. Without a doubt it is the best place in the area from which to observe migrating birds. We also examine the climate of the Campo de Gibraltar to be able to understand why it has such unusual flora and fauna. In *Chapter II: Los Alcornocales: a resting area*, we describe the flora and vegetation of this park, a stopover for the majority of migrating birds, comprising some superb cork oak and Mirbeck's oak forests containing unique, relict ferns. In *Chapter III: The Rock of Gibraltar*, we try to present a broader and fairer view of the migratory phenomenon in the Strait, because most works up to now have described migration either only from Gibraltar or only from the Spanish coast of the Strait, and never as the complex whole which it really is. This chapter was written by the current director of the Botanic Garden of Gibraltar, and a leading member of The Gibraltar Ornithological & Natural History Society (GONHS), Dr John Cortés, a particularly good friend of mine for many years now. In *Chapter IV: Migration in general and special characteristics of the Strait*, we analyse the different theories of why, how, where and when migration takes place, while not forgetting either the birds' flight or modern monitoring techniques and their repercussion on the Internet. Then, in *Chapter V: The migration of the White Stork*, we take a look at one of the most spectacular migrations, analyse its strategies and the problems faced when crossing the channel, as well as their stopover and roosting places. The aspect of migration most appreciated by ornithologists who come to the area and who want to see as many raptors as possible, and the closer the better, is the subject of *Chapter VI: Raptors and other migrants in the Strait*. We examine the flight strategies of a total of 28 species, and their most outstanding features. We must not forget the word of warning in *Chapter VII: Misadventures and deaths*, which deals with the problems and dangers faced by migrating birds and then, finally, in *Chapter VIII: The ornithologist's and traveller's guide*, in which we advise those who wish to indulge their passion for birds, and who have often come a very long way, where to get a front row seat so that they don't miss a single migrant species crossing the sky, either in the pre- or postnuptial period. There will be a brief description of the area known as the Campo de Gibraltar including its towns and the nearest nature enclaves for the visitor to enjoy. We include maps, routes and places of faunal and floral interest in the area as well as addresses of local ornithological societies.

Preamble

Geographical location

I should like to warn readers that, far from wishing to put your geographical knowledge in doubt, I feel bound to dedicate a few lines to describe the exact location of the events you are about to read of : it is no other than the famous Strait of Gibraltar, and, to be more precise, the European side of it. It is situated at the extreme south western tip of Europe between the Mediterranean Sea and the Atlantic Ocean at a pivotal point between the continents of Europe and Africa. It was here that in Greek mythology

Geographical location of the Strait of Gibraltar :

Hercules used his fabulous strength to separate Europe from Africa by leaning on the two famous pillars (Gibraltar and Monte Hacho). The Arabs called it Bahr-z-zohak or narrow sea.

Research studies in the Strait

Since the XIX century many distinguished foreign ornithologists have spent time in these tempestuous southern lands, contributing a rich and well documented vision of ornithology in the area. Among the most famous are L. Howard Irby (*The Ornithology of the Strait of Gibraltar*), Willoughby Verner (*My Life Among the Wild Birds in Spain*), H. Saunders (*Catalogue des Oiseaux du Midi de l'Espagne*), and the first ornithologist in Europe to observe the Red-rumped Swallow in the area of La Janda, Sir G. Lathbury (*A Review of the Birds of Gibraltar*) and P. Brudenell-Bruce. The latter was the first ornithologist to observe the White-rumped Swift (*Apus caffer*), in 1964, which he had initially confused with a Little Swift (*Apus affinis*), (Brudenell-Bruce 1966). The first Spaniard to write in detail about the area was Pedro López de Ayala who published his *Libro de la Caza de las Aves* in the XIV century.

After WWII and especially from the fifties on, coinciding with the first important discoveries about migration, the Strait began to feature significantly in studies in bird migration, mainly in research and field notes: Moreau & Moreau, 1956; Mountfort, 1958; Bruhn, 1958;

Opposite

In the foreground eroded rocks on a beach in the Natural Park of the Strait, with the shores of Africa in the background.

Feeny, 1960; Henty, 1961; Fry, 1961; Bernis, 1963; Casement, 1963; Colsten & Cowlen, 1963; Evans, 1967; Lathbury, 1968; and Pineau & Giraud-Audine, 1974-1976. In Gibraltar several works on local and migratory ornithology were published (Lathbury, 1970; Evans & Lathbury, 1973; García, 1973; Finlayson et al., 1976).

In spring 1972, during the university Easter holidays, Professors Francisco Bernis and Manuel Fernández Cruz from the Vertebrates Department of the Complutense University of Madrid arrived in the Campo de Gibraltar to get acquainted with the area, with a view to begin a study of migration in the Strait the following autumn. They had come in Bernis' car and decided that Fernández Cruz would stay at the Mirador del Estrecho while he would carry on to Algeciras. When Bernis returned to the

Sir Gerald Lathbury, governor of Gibraltar in the 1960's, who encouraged local interest in ornithology and was a pioneer of studies in migration on the Rock.

Mirador he asked Fernández Cruz if he had seen anything, to which the latter replied saying, yes, he had seen a dozen short-toed eagles arriving. After a stunned silence, Bernis said: "Manolo, you're pulling my leg". Fernández Cruz answered back saying that if he didn't believe him, he should take him to Algeciras station to catch the evening train back to Madrid. A few days later Joaquín Araújo and Olegario del Junco, founding members of GEMRA (The Spanish Bird of Prey Migration Association) came down to see for themselves.

This little story is a true reflection of how little was known, in the early seventies, about migration. In 1997, during their prenuptial passage on an upward thermal, I saw 101 short-toed eagles – something which nobody would have believed years before.

In 1971 Bernis had heard that a group of foreign ornithologists were trying to obtain permits from the Spanish Foreign Office so as to be able to carry out important research into bird migration in the Strait of Gibraltar. Very annoyed, he pulled a few strings, these permits were denied and he got a scholarship from the Scientific Research Institute (CSIC) and the Department of Zoology of the Complutense University in Madrid, thanks to which he was able to start his own research.

In the summer of 1972, led by Professor Francisco Bernis, some students, graduates and members of the SEO arrived in the Strait of Gibraltar, in order to start the first campaign study in the Strait into postnuptial migration. Among these first volunteers were some now very well-known people: Joaquín Araújo, José Luis Tellería, Jesús Garzón, Fernando Parra, Manuel Fernández Cruz, José Luis Pérez Chiscano, Ramón Elósegui – about thirty altogether

but too numerous to mention, all ready to spend some arduous but unforgettable days in the Strait. These ornithologists had to put up with strong winds, the burning, Andalusian summer sun and even some harassment by the military authorities in the area, as when Ramón Elósegui was briefly arrested by some soldiers from a coastal detachment, who thought someone near the coast looking at military targets through binoculars was a highly suspicious individual. I was also fortunate to spend that summer with these young bird enthusiasts and since then I have never missed a date with either pre- or postnuptial migration!

The campaigns in 1972 and 1974 were made possible thanks to funds made available by the Science Division of CSIC and the José Acosta Institute of Zoology and the results were published in volume 21 of *Ardeola*, a magazine edited by the SEO. In the same volume, José Luis López Gordo published an article about the migration of the Bee-eater. Bernis wrote a series of long articles for the same magazine between 1972 and 1974. In 1975 there was no campaign due to lack of funding, but in 1976 and 1977 the Juan March Foundation provided the necessary funds and research continued.

Many pieces of work have been published as a result of these campaigns and other research is being carried out. The most extensive, and pioneering, work is *Bird migration in the Strait of Gibraltar*, which comprises two volumes: *Volume I: Soaring Birds* written by Francisco Bernis, and *Volume II: Non-soaring birds*, by José Luis Tellería, whose doctoral thesis it was. We consider them to be essential works of reference for any ornithologist interested in migration, both from historical and strictly scientific points of view.

In 1985 a study by Manuel Fernández Cruz,

funded by the Ministry of Public Works, looked at the postnuptial migration of the White Stork, in which 35,000 individuals were counted. In the 90s a census in a similar study gave an incredible 113,000 birds.

In 1986 the German ornithologist Gudrun Hilgerloh asked the Spanish military authorities in the Strait if she could use their radar apparatus to study the nocturnal migration of passerines. They refused to help but the British Admiralty in Gibraltar were more cooperative and in her research she established that many of these little birds fly at night and at altitudes of more than 1500 metres.

One of Gudrun Hilgerloh's compatriots, Joaquin Griesinger, studied the postnuptial migration of Griffon Vultures which were ringed as chicks in Navarra in 1990. They were marked with paint and tiny transmitters were attached in order to monitor them. At times observers of them were in radio contact from both shores of the Strait. This was a new experience because, up till then, monitoring of migrant birds had been carried out either from the European side or from the African but not both sides at the same time. Due to lack of time the work could not be finalized, but about 2,000 Griffon Vultures had been registered by the third week in November and we know now that these carrion birds are still crossing the Strait in December.

In 1992 a book was published by Gibraltarian Clive Finlayson, *Birds of the Strait of Gibraltar*, outlining his perspective of migration (using Bernis' and Tellería's studies as a base) and supplying hitherto unpublished data on the migration of sea birds, observed easily from the Gibraltar lighthouse.

Studies of the migratory passage of the Black Stork began in 1993 with a group of volunteers under the guidance of César Sansegundo. However, he had to leave the area and direct operations from a distance. So, as he admits in the 102nd number of *Quercus* magazine, "on few occasions have we been in two places at once", so the rigorousness of this scientific work must be called into question.

That was also the year when a new migratory study of the white stork was begun, financed by the Complutense University of Madrid and directed by Manuel Fernández Cruz. Tarifa Town Council collaborated in this project by providing an old school on La Peña, used in spring by the army in its horse breeding programme. The project was part of a course on vertebrates which has been running for seven years now. At the moment the premises are occupied by COCN (the Black Stork Ornithological Collective) who have signed an agreement with Tarifa Town Council. In that first year 70,000 White Storks were counted, a figure which was easily surpassed in the years that followed, reaching 120,000, probably due to an increase in the Iberian population and more effective counting methods.

In the first bulletin published by the Migres Programme (Pilot project 1997) it says: *In the context of an Interreg II programme of the European Union for cross-border cooperation with non-Member states, the Andalusian Regional Government is promoting and financing various activities relating to bird migration across the Strait of Gibraltar. Among them is a study, directed by the CSIC at the Doñana Biological Station, into its socio-economic potential as a generating source of wealth in the Area of the Campo de Gibraltar, in particular, and in the rest of Andalusia in general; there is also a project, developed by EGMASA to construct new, and adapt existing, infrastructures*

designed to channel the public interest generated by bird watching; and then there is the Migres Programme which monitors bird migration in the Strait of Gibraltar and is coordinated by the Spanish Ornithological Association (SEO)." The first phase, from 1997-2001 was carried out using the ideal method for continued long-term research, but from 2002 on and as long as funding is available, the "constant effort" method has been chosen,

by which *"the recording of data is objective and standardized and data can be compared over the years".*

2003 saw the publishing of the *Field Guide to the Birds of the Strait of Gibraltar*, by David Barros Cardona and David Ríos Esteban, a Spanish and English bilingual edition in which there is a comprehensive review of all the birdlife of the Strait with detailed information regarding habi-

Francisco Bernis Madrazo, the internationally renowned ornithologist, who pioneered studies in bird migration across the Strait in the 1970's

tats, migration, species distribution maps and places of ornithological interest. It is an essential book for visitors to the area, for those who want to be informed or for those who just want a good book on ornithology. The authors put its publication down "to the lack of a guide book on birds concerning one of the most important and prestigious places in the world for observing migration". For our part, and without wishing to appear too chauvinistic, we are immensely proud that two local ornithologists have put together a piece of work of such magnificent quality and so well documented.

Finally, although it is not really a study as such, I must mention a masterful book of photographs which came out in March 2004 and is signed by our friend and brilliant nature photographer Manuel Castro Rodríguez – or Manolo Castro as he likes to be called. Its title *From the Alcornocales to the Strait (The naturalist's view)* tells us what to expect. His intelligent camera and singular eye for beauty show us all the biotopes in the Alcornocales Natural Park and in the recently created Natural Park of the Strait. From the 300 magnificent photographs we can witness how, just a few kilometres away from the bustling cities, there is still a world of nature waiting to be discovered even if we go without the magic camera of this amazing photographer.

As I correct the proofs of this book a marvellous work has just been published called *Los Alcornocales Parque Natural* written by Joaquín Araújo, Emilio Blanco and Carlos Santos. The photographs are by Juan Tébar a native of Cádiz and a very good friend of mine. While not wanting to appear biased I must say that his impeccable images take the reader into a magical world where all that is missing to give one that feeling of total immersion in the depths of Nature are the unique fragrances of this park.

It would not be fair of us to forget to mention those people from the Campo de Gibraltar, either by birth or adoption, who, unselfishly or anonymously, have worked and are working hard to divulge, study or simply show to their neighbours the wonders of this corner of the country which we are fortunate to share, along with relict ferns, White-rumped Swifts, Short-toed Eagles and multi-coloured meadows. These are distinguished people and well-loved by those of us who have a passion for our nature and our land, especially when we see they love it as much as we do.

One of these illustrious people was Paz Lerma (the former Duchess of Lerma whose full name was Doña Paz Fernández de Córdoba), a conservation pioneer, described by the writer Juan José Téllez as "the ecologist duchess". She owned several splendid large estates in the area and was always aware and respectful of the biodiversity in them. She was the author of a book of photographs entitled *Algunas de mis fotografías*, a copy of which she kindly dedicated and gave to me. During the last years of her life she banned deer hunting on one of her estates, El Jautor, near Alcalá de los Gazules.

The renowned British botanist, Betty Molesworth Allen, was famous in the sixties for discovering the fern *Psilotum nudum* in the Sierras of Algeciras, and other major ones like *Christella dentate, Diplazium caudatum* and *Pteris incompleta* or *Arisarum proboscideum*, a plant of the aroid family. She was married to Geoffrey Allen, a daring nature photographer and together they travelled the five continents, before settling in the quiet town of Los Barrios. It was thanks to her, and one of Mr Allen's photos of a White-

Opposite

Jorge Bartolomé, ornithologist from the Complutense University of Madrid during a census of migrating white storks near Gibraltar in the 1990's.

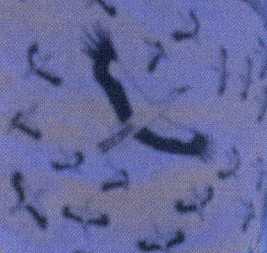

rumped Swift leaving its nest in the Sierra de la Plata, that the Campo de Gibraltar and the Alcornocales gathered worldwide fame as an enclave of important flora and fauna.

The first nature photographer I met was Lucas Millán Millán. A native of Algeciras, where he had a fabric shop, he had some incredible photographic equipment including a Rollei 6x6 format camera, with which he took excellent photographs: fungi, prehistoric cave paintings and scenes from the Campo de Gibraltar. Being extremely modest he was not well-known but was happy taking photos and showing them to his closest friends. When I told him one day I was going to give a talk in the town of Palmones on flamingos, he said I was a brave man as he would be incapable of showing photos to so many people. His photographs had personality... what people now call style, although he used to play it down and said he was just portraying what he saw.

Back in the eighties I heard that a school teacher had come to live near Jimena de la Frontera, in the Campo de Gibraltar and he was rumoured to be tearing up every rare species of plant he came across, probably to sell to foreign herbalists. And he wasn't from round these parts either. Luis Federico Sánchez Tundidor, or plain Federico, is one of those illustrious botanists who knows every square inch of flora in the Campo de Gibraltar and is a real authority on botany. He has published a great deal on the subject and his extrovert character is easily recognizable in any symposium and congress on the Campo de Gibraltar.

Domingo Mariscal Rivero, another teacher and country colleague of Federico's, with whom he has often worked, is one of the most versatile personalities we have. Both in his profession and spare time he teaches people about the flora of Los Alcornocales, gives talks, attends conferences and writes in local magazines like *Almoraima* and *Alimoche*, among others. He takes an interest in anything to do with culture in the open air, from a *Catologue of pteridologic flora under threat in the south of the Alcornocales Natural Park* to discovering prehistoric settlements and defending them publicly from the world of cement.

The English –Andalusian Cristina Parkes, an admirer, like myself, of birds of prey, got me in to several large estates in the area. We trekked up and down many *sierras* and studied the Eagle Owls on the Aciscar estate for many years. Her current passion is the Black Stork and she has written about it and our area in many magazines and on the internet in both English and Spanish.

In the eighties, just after the Spanish border with Gibraltar was opened, I was once out in the country in an area called Las Corzas, near Algeciras, looking for salamander larvae to photograph when, at a pond, I bumped into a couple of *foreigners*, whom I greeted briefly, picked up what I wanted and hurried off to do something else. Mr John Cortés and I have never forgotten that chance meeting by that pond. He is one of the *yanitos* (a name people from the Campo de Gibraltar give to those from the Rock) and a pioneer in fostering the study and conservation of wildlife and the natural environment – at first just his own, but now ours too. He is Director of the Botanic Garden of Gibraltar and member of GONHS (The Gibraltar Ornithological and Natural History Society). The chapter in this book devoted to the Rock of Gibraltar was in fact written by him and will no doubt reveal to most readers many impressive and hitherto unknown details of nature, unrelated to the ever topical Gibraltar apes.

Alfonso González Carbonell, from La Línea de la Concepción, and currently its local councillor for the environment, is one of the most knowledgeable authorities on small birds in the whole area. A founder member and, in the eighties, president, of GOES, an ornithological group dedicated mainly to the scientific ringing of birds, he has had both in his hands and in invisible nets all of the main species of birds which have flown over the area.

Lothar Bergmann, a very likeable Andalusian born in Germany, is a person who has specialized in promoting Southern Art (*Arte sureña*) – a series of caves and shelters containing parietal art, and, if not the most then one of the most valuable prehistoric elements in the area. We are indebted to him for the preservation of the drawings in the Moro cave, a jewel from the Upper Palaeolithic, discoveries of shelters with paintings and the publicity given to all these art treasures.

A married couple – José Ramón Sogrob and Mari Carmen Fajardo - is responsible for the growing interest in fungi in the Campo de Gibraltar since the nineties. Their articles, talks, posters and much, much more have helped foster this interest and their converted *aficionados* are rapidly becoming real gourmets.

And finally let me mention Pablo Ortega whom I encouraged to watch and, later, photograph birds. He is a co-founder of GEODE (the Association of Ornithological Studies in the Strait) together with Martín Caballero, Juan Carlos Castro, Rafael Ruiz and Ana Juárez, and is the only ornithologist in the area whom I have come across in the Strait, year after year since the eighties, watching those tired birds arriving during their prenuptial migration.

I know that one or two people may feel hurt at not being mentioned along with these illustrious names, but I have preferred to run that risk rather than omit those who have been and still are key figures in the nature world in our area. I must apologize to whoever may feel offended at being forgotten, but on the other hand I don't mind too much because it's a sign that the person is still young and with a great future ahead! I would also just like to thank all those anonymous ornithologists who come from other regions of Spain and from countries such as the United Kingdom, France, Germany, Holland, Portugal, Hungary, Finland etc, who help to spread the word about this unique migratory enclave, while sharing a hobby or indeed, in some cases, a profession which gives us so much satisfaction and the opportunity to make friends.

Before closing this chapter, it would not be fair of me to ignore an organization which has been carrying out great cultural work in the area for years: the Cultural Department of the Association of Townships of the Campo de Gibraltar, which organizes the prestigious Conference on Flora and Fauna in the Campo de Gibraltar and publishes its findings in its own magazine, *Almoraima*. Another important body within the Association is the Instituto de Estudios Campogibraltareños and its Section X of Medicine, Biology, Ecology and Natural Sciences. I shall not name any particular individuals because these have changed over the years but our thanks are due to them for helping to promote and make our patrimony available for everyone's enjoyment rather than keep quiet about it!

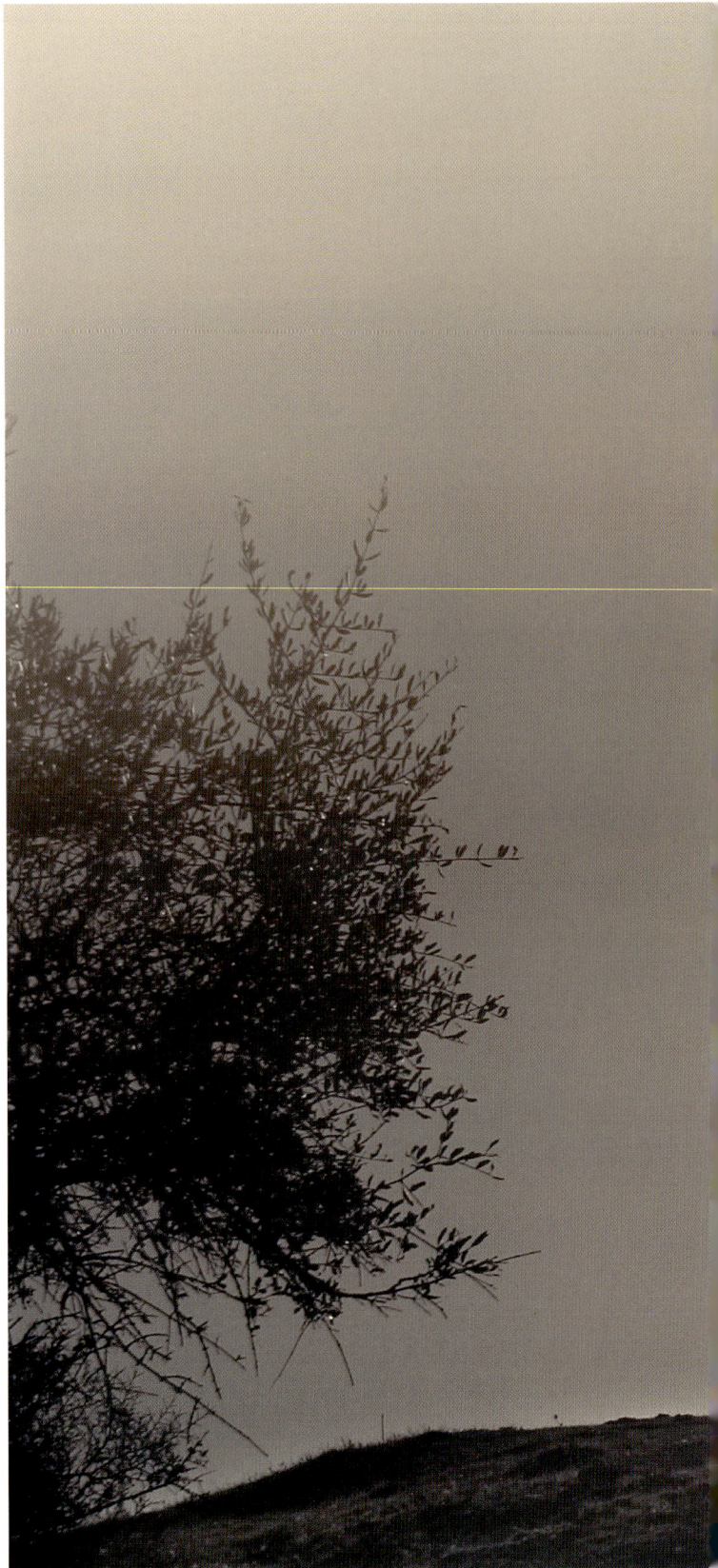

In the observatory at El Algarrobo early-rising ornithologists awaiting the arrival of Honey Buzzards on a hot, September day in 2006.

The formidable migration of soaring birds across the Strait of Gibraltar is one of the most breathtaking natural events still to be witnessed in our world.

Francisco Bernis
Bird migration in the Strait of Gibraltar.
Vol. I Soaring birds, 1978

The Natural Park of the Strait

On the 4th March 2003, in what was a bold ecological and political move, the Andalusian Regional Government announced the creation of the Natural Park of the Strait. This success was achieved in the face of scheming opposition from the local government of the town of Tarifa, which tried to convince the local population that the park would put a brake on the development of tourism in the area, while hiding the fact that it would be the same old story for those benefiting from the absence of the park: the few landowners who would speculate with their land, just as they did in Zahara de los Atunes and Atlanterra, where one of the most the scandalous affronts to urban development in the last few years has taken place. As a perfect example of such excesses and under the umbrella of professional speculators in the construction world, what you can see there now are enormous blocks of cement clustered together and completely hiding from view the immense beach of white sand and the magnificent sunsets over the Natural Park of La Breña y Marisma del Barbate. To add insult to injury there is no suitable sewage disposal system to cope with the holiday season population, nor is there a reservoir which guarantees fresh water supply, so they have to put up with water rationing, more typical of the Sahara than of a modern European residential area.

Surface area and geographical limits

On the 9th February 1999 the Andalusian Regional Government approved the framework for the Plan for Natural Resources on the Coast between Algeciras and Tarifa and on the 22nd April of that year it was agreed to include in that designated area the Cerros del Estrecho and some land at the southern tip of the Natural Park of

In the foreground the pebble beach of Guadalmesí with Jebel Musa in Morocco in the background.

Los Alcornocales and part of the sea, the Getares inlet and the Punta de San García, beyond Punta Carnero.

The Natural Park of the Strait has a surface area of 19,126 hectares, of which 9,880 are on land and the remaining 9,246 sea, a nautical mile out along the coast. It is divided into two sectors, the Atlantic and the Mediterranean. The former covers the *sierras* of Betis, San Bartolomé and La Plata and the land between the N-340 road and the littoral. A dual carriageway, the A-48, is now being built parallel to the N-340 road. It will link the Bays of Gibraltar and Cádiz and be a boost to tourism and industrial development in those two areas The A-48 road limits the eastern sector to the north, from Punta de San García in the east (at the residential development of the same name, belonging to Algeciras), following the coastline as far as Punta Camorra, near the D-8 Military Coastal Battery belonging to the Ministry of Defence. It is here where the rough, stormy waters of the Strait pound the shore and where the decline in fauna, flora and, in some places, the landscape, is sadly noticeable.

From the end of the fifties until 1974 yours truly was mad about underwater fishing. This was a virgin coast until the end of the sixties, so that those of us who were privileged enough to scubadive among Poseidon's meadows or over immense carpets of algae, remember those waters as if they were huge natural fish farms, where the numbers of fish we saw remind us now of documentaries you see on television about exotic seas teeming with multicoloured fish life. At that time nobody could have guessed that this wonderful exuberance of nature was condemned to death.

With some goggles, breathing tube, fins and a sweater (to avoid scratching myself on rocks) and an almost fundamentalist approach to marine fauna when I was 15 I would dive into a fascinating world near beaches, rocky reefs and coves. I can remember enormous shoals of rubberlip grunts, bream, sea bass and bogues, and others, swimming next to me or, in the case of the bogues, all around me at a safe distance.

On one memorable occasion after several days of strong hurricane force levante winds, I went fishing at a place called Arenillas. My father was keeping an eye on me from the beach and we had agreed I would swim parallel to it because I was still quite young and inexperienced for such a dangerous place as the Strait. I was swimming at a familiar spot over a white, sandy seabed, where I usually caught sole (*Solea solea*). It was about four metres to the bottom, but because of the groundswell visibility was only good from about a metre under the surface. Looking down I could see the bottom moving, like an enormous rippling carpet, with the rhythm of the waves. I thought it was algae that the storm had ripped up and, as often happens, they were now covering the sea bottom and, therefore, hiding the motionless sole from my now expert eyes. As the algae covered a huge area and I was bored swimming on the surface I decided to dive right down to have a closer look. I took a deep breath, filled my lungs and down I went. No sooner had I gone through that milky-like layer which was spoiling my vision than I was presented with one of the most amazing sights I have ever beheld in the waters of the Strait. As far as the eye could see, in all directions the seabed was literally carpeted with millions of red mullet (*Mullus sp.*), packed together and moving in time to the groundswell. I yanked my head out of the water and spluttering salty sea water I shouted: "Dad, dad, they're not algae, they're red mullet!" There seemed to be millions of fish over a vast area, motionless unless you touched them.

Do not worry, dear reader, I am not going to tell you many more of my little stories. I am just giving you an example so that you can compare what there used to be, just a drop in the sea of geological time ago, and what we have left today. In no more than a decade, the fauna in the Strait and that of most of the seas on Earth, has vanished because of pollution and human greed in the form of overfishing.

The park in question covers a surface area of just a couple of hundred hectares, but never has such a small space contained such an enormous quantity of life, since the majority of those 300 million birds which cross the Strait annually do so through its sky, and in the course of a year several million migratory birds stop for rest, food or shelter on the hillsides which slope down to the sea like stony, green-, yellow or ochre- coloured waves, depending on which season it is.

The morphology of the littoral

The Mediterranean side is the most rugged and has only two white sandy beaches, Getares and Cala Arena, a cove which is much smaller and fortunately is not accessible by vehicle. This stretch of coast, apart from that, is uneven, with rocks and rocky reefs which make bathing awkward from the beach although they are ideal for those who go diving for shellfish. This very harsh morphology is due to the fact that part of this littoral is made up of the foothills of the Cabrito and Bujeo *sierras*, which are perpendicular to the coast and right next to it.

In this sector there are a series of hills cut through by streams which run into the Strait. The hills are rounded with steep slopes and signs of erosion due to the action of the levante

winds on a terrain covered in very little shrub land. Although they do not rise to great heights their proximity to the sea makes them look very uneven and rugged. Apart from erecting beacon towers, possibly to detect marauding corsairs, in the past the look-outs set fire to the land to be able to detect these more easily. The scarcity of trees may also be due to the action of the wind and the landowner-farmers with cattle on the land as their only viable source of income.

The outstanding landmark in this sector is the cliff at Punta Carnero on which the lighthouse stands. From the western end of Getares beach, where there is a military bunker, to Punta Secreta a series of small cliffs alternate with rocky beaches and reefs. The area around the Punta Carnero lighthouse is spectacular to the eye due to the sandstone and marls which constitute the rocky platform. As the marls are less resistant to erosion the deep cavities and holes in the sandstone blocks give it a honeycomb effect.

On the left hand side coming up to the lighthouse just before the wide bend to the left there is a curious geological formation, namely some bare strata of flute cast. These sediments are formed on the seabed when sandstone deposits pile up on mud and eventually appear on the earth's surface like a photographic negative. On the track leading from Palancar to Monte Ahumada there is also an interesting fossil fragment which was originally on the seabed.

These beaches with reefs or slabs of rock are ideal spots for certain species of shorebirds which winter here or rest during migration. There are also a total of eleven beacon towers and several military bunkers on this stretch of coast. The beacons were built in the XVII century and later used as rather inefficient coastal batteries, since, being so far apart, they couldn't do the job they were built to do, which was to defend the coast from invaders. (Sáez, A., 1987 *Aproximación a las torres Almenaras de la bahía de Algeciras)*. Now obsolete, the artillery batteries on the coast of the Strait date back to the end of the Civil War, when they were built to guard against an Allied invasion, or at least as a deterrent. Some have been remodelled and moved inland to be equipped with modern armaments and technology.

From Guadiaro to Conil there are 478 military bunkers, very close to the littoral and it was the Military Junta of Burgos which had them built (Escuadra, *et al.*, 2003, *Almoraima* 29). They formed part of a national plan to seal the borders of the Mediterranean, Pyrenees, Strait of Gibraltar and the Balearic and Canary Islands, forming part of the Jevenois Commission. These fortifications, then, are scattered along the coast and used by different fauna, sometimes even by drug traffickers and illegal immigrants. As nearly all the coast has been or still is a military zone, there are very few built-up areas in the park, except for the developments in the east (San García, Getares and the Punta Carnero lighthouse), the mouth of the river Guadalmesí, also to the east, and to the west of the town of Tarifa there are those at Bolonia and Cape Gracia, as well as farmsteads dotted along the coast.

As most of the area was declared a military zone, after the war no civil construction was allowed. To connect up the coastal batteries a series of unpaved roads or tracks were made just behind the hills next to the sea, so that military vehicle movements could not be detected from Gibraltar by allied forces. There was only one place where it was impossible to hide part of a track from potential enemy eyes, so at La Pantalla, a reinforced concrete wall was built, long

Natural Park of Strait of Gibraltar :

- terrestrial park
- maritime park

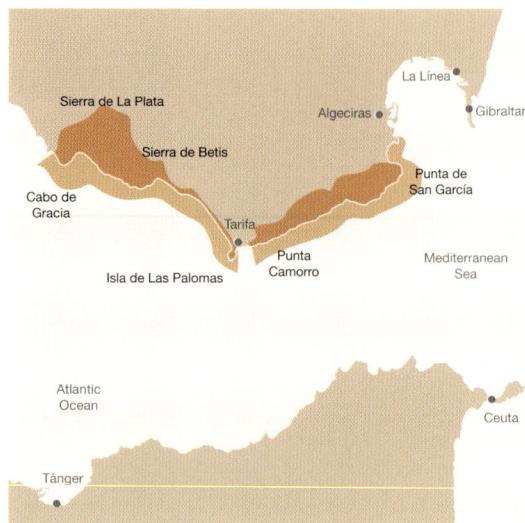

and high enough to hide lorries and artillery, and shaped like rocks. As with the bunkers, there is an interesting collection of fauna in this unusual place. In theory you still need a permit from the Military Governor of the Campo de Gibraltar to go along these tracks or you may be fined. Each track has a No Entry sign where it starts. Fortunately common sense prevails and the military authorities are indulgent.

Between the Punta Carnero lighthouse and Tarifa there are a few pebbly beaches, which are

Opposite

Watching marine mammals is one of the favourite activities for tourists visiting the coastal towns of the Campo de Gibraltar.

Sunset at the Punta Carnero lighthouse in Algeciras. Built at the entrance to the Bay, new technologies have made it almost obsolete.

very unusual on the Cadiz coast. These are small coves where the sea beats against the cliffs and tears off stone, which through erosion is gradually transformed into pebbles on the Arenillas and Guadalmesí beaches.

Within the park there are two nature reserves, both in the Atlantic sector: the Nature Area "Playa de los Lances" and the Bolonia Dune Natural Monument, which we will deal with shortly. This area has some wonderful, wide, white sandy beaches. One stretches 8 kilometres from Los Lances on the island of Las Palomas in Tarifa right to Punta Paloma.

The Nature Area "Playa de los Lances" is an ornithological reserve set up by the local government in 1987, when it won the Ford Prize for Conservation. The original idea was from the GEODE (Association of Ornithological Studies in the Strait) of which I was the president. Based in Algeciras it was a pioneering organization in the ringing of raptors in the area and has now become known as AGADEN, ecological in outlook. Anyway, the Town Council

Opposite

Kentish Plover, an endangered species because of increasing human presence on beaches, one of the few shorebirds still nesting in the sand dunes.

of Tarifa was informed about how important the area of its littoral is for migratory birds and the need for its protection. The Council therefore declared the "Playa de los Lances" an ornithological reserve. Nowadays, the only great advantage obtained as a result of this declaration is the ban on building in the area next to the football field, because otherwise the place is full of stray dogs, groups of people on horseback, noisy quad motorbikes trying to emulate the Paris – Dakar race and other pleasure seekers contravening the National Coast Law. This is the part of coast where most wind and kite surfers congregate because of the wide open spaces and the strong winds, both *poniente* and *levante*. The wind, which until the sixties had held back tourist - and therefore economic – development in the municipality of Tarifa, is now the driving force which attracts visitors from Spain and all over Europe to make it the most popular place in the area.

Inside the park a wide plain sees the joining together of the Jara and Vega rivers and the Salado stream. The plain is flanked by *sierras*: Cabrito in the east, Ojén in the north and Fates in the west. The mouth of these rivers creates a floodable area protected by the sandy beach. This lagoon is one of the four coastal wetlands next to the Strait and apart from this one we have the Nature Area of the Guadiaro Estuary, the Nature area of the Wetlands of the River Palmones and the Torre Guadiaro lagoon. Both local birds and migratory birds when they cross the Strait find food and refuge in these four areas: a total of 228 different bird species have been counted there altogether (López, 2000). They are closely linked, of course, to the wetlands on the northern Atlantic coast of Morocco, but, sadly, these too are declining.

The Enmedio *sierra* tumbles right down to the sea and, as its name suggests, splits up the beaches into two, the Los Lances beach and the Valdevaqueros cove. The La Peña campsite and a beacon tower are here, on the right of the N-340 going towards Cádiz.

At the Valdevaqueros cove, where the river Valle runs into the sea and the interaction of tide and river water form a lagoon at its mouth, there is a magnificent sand dune, which unfortunately has been deteriorating since the eighties, due to being constantly trampled by tourists hiking to the top. The spectacular increase in the number of tourists in Tarifa and the surrounding area has accelerated this deterioration and consequent spreading of the dune into a pine plantation, which was originally introduced to check the dune's advance. There is no type of protection whatsoever, possibly because the Department of Coasts of the Ministry of the Environment does not know how to stop people damaging the dune. Until the eighties there used to be lines of strong canes on the crest which stopped or slowed down the accumulation of sand on the leeward side but they were broken down by people crossing to the windward side and thus the sand started spreading further afield. As for the river Valle, it flows in a north–south direction between the Fates and San Bartolomé *sierras* forming a valley which, as we shall see, is of major importance in the postnuptial migration of soaring birds.

Along this rugged littoral, just after the Valdevaqueros cove we have Punta Paloma, a promontory of some 444 metres jutting into the sea from the San Bartolomé *sierra* and forming, as it does, the western end of the Valdevaqueros cove and the eastern end of Bolonia beach. This huge sandstone block looks as if it is floating on other geological deposits. In the surrounding area, especially on the west side you can see some large rocks which have tumbled down from the mountains in previous eras. The land then slopes down gently to the cove of Bolonia. Here we have the Bolonia Dune which rises up between the Roman ruins of Baelo Claudia and the military zone at Punta Camarinal – which takes its name from the plant called Portuguese crowberry (*Corema album*) growing here at its southernmost point. Thanks to the *levante* winds sand is carried in a north-westerly direction and piles up to form the dune. However as tourists are constantly hiking up and down it, it is losing height and, as with the one at Valdevaqueros it is threatening to invade the pine plantation put there to stop its progress.

The beach at Bolonia, together with the Roman ruins of Baelo Claudia – where there was a major fish salting industry, is a dream come true for tourists at any time of the year. As with the beaches at Los Lances and Valdevaqueros the sand which builds up where streams run into the sea has formed a small lagoon and attracts many shorebirds during their annual migration.

At the end of the beach stands Cape Camarinal, a portion of limestone stuck on to the end of the La Plata *sierra*. This is where the Romans quarried the rock they needed to build Baelo Claudia. Today you can still see huge chunks of rock cut and ready to be transported. Beyond this cape there is another, Cape Gracia, also known as Torre Gracia, which marks the end of the Natural Park.

The park has two distinct littorals, the eastern or Mediterranean and the western or Atlantic littoral. The morphology of both coasts depends on the influence of the water masses.

The waves from the Mediterranean on the eastern side do not have sufficient space to become really significant and therefore the erosion co-efficient is quite moderate. On the other hand, as the Atlantic Ocean is so great, the action of the wind causes enormous waves to arrive at the coast unhindered and with considerable erosive power due to both their speed and height. Erosion has led to the deterioration of the coastline and the formation of large, sandy beaches.

A study of the morphology of both stretches of coast shows clear differences between them. The role of the tides in the Strait is also important as they affect both the morphology and the marine fauna.

Because the Mediterranean Sea is geologically quite small, more water evaporates there than is brought in by large rivers like the Danube and the Nile. This means its level is lower than the Atlantic and therefore water flows from the latter into the Mediterranean via the Strait of Gibraltar. This colossal amount of water on the surface creates coastal cross currents, much feared in the area by sailors and fishermen. These torrents of water within the sea have tremendous force, but coastal predator fish like grouper, croaker and sea bass are strong swimmers and easily capture the little fish which really struggle in such conditions.

Scattered along many parts of the coastline, excluding the areas of sandy beach, you can find some very hard, abrasive rocks which are home to a special type of fauna, and even flora, whose lives are spent partly in and partly out of the sea. Goose barnacles, limpets, sea anemones and many other creatures are able to withstand long periods with little contact with water and long exposure to the sun, although they are constantly threatened by dehydration.

The climate in the Campo de Gibraltar

The following description of the climate is valid for the whole of the Campo de Gibraltar area, given that it includes this park. This is to avoid unnecessary repetition when we deal with the Alcornocales Natural Park, when only the odd exception will be mentioned.

The Strait of Gibraltar lies between the Mediterranean and the Atlantic Ocean. It is at the confluence of three marine areas: the Lusitanian, the Mauritanean and the Mediterranean. This is evident both from the type of flora and climate and also from its marine fauna, which, although not dealt with in this book, is made up of living creatures from these three areas.

A semi-humid Mediterranean climate predominates, with mild winters typical of the Gulf of Cádiz, and is determined by four factors: the subtropical anticyclone from the North Atlantic (the Azores anticyclone), the sub Polar cyclone from the North Atlantic (Iceland), the Saharan thermal depression and the Peninsular thermal centre.

The position of the Azores anticyclone is a determinant factor for the climate in Europe and North Africa and it affects the arrival of *poniente* or *levante* winds and cycles of rain or drought. If its centre is at the 30º N parallel then moist winds come to the Peninsular, bringing abundant rainfall. When it is at the 40º N parallel (on a level with Madrid) troughs of low pressure do not arrive at the peninsular and periods of drought follow. Finally, if an anticyclone

In the Strait, the average wind speed is 22 Km/h every month of the year and, also every month, there are gusts of more than 109Km/h.

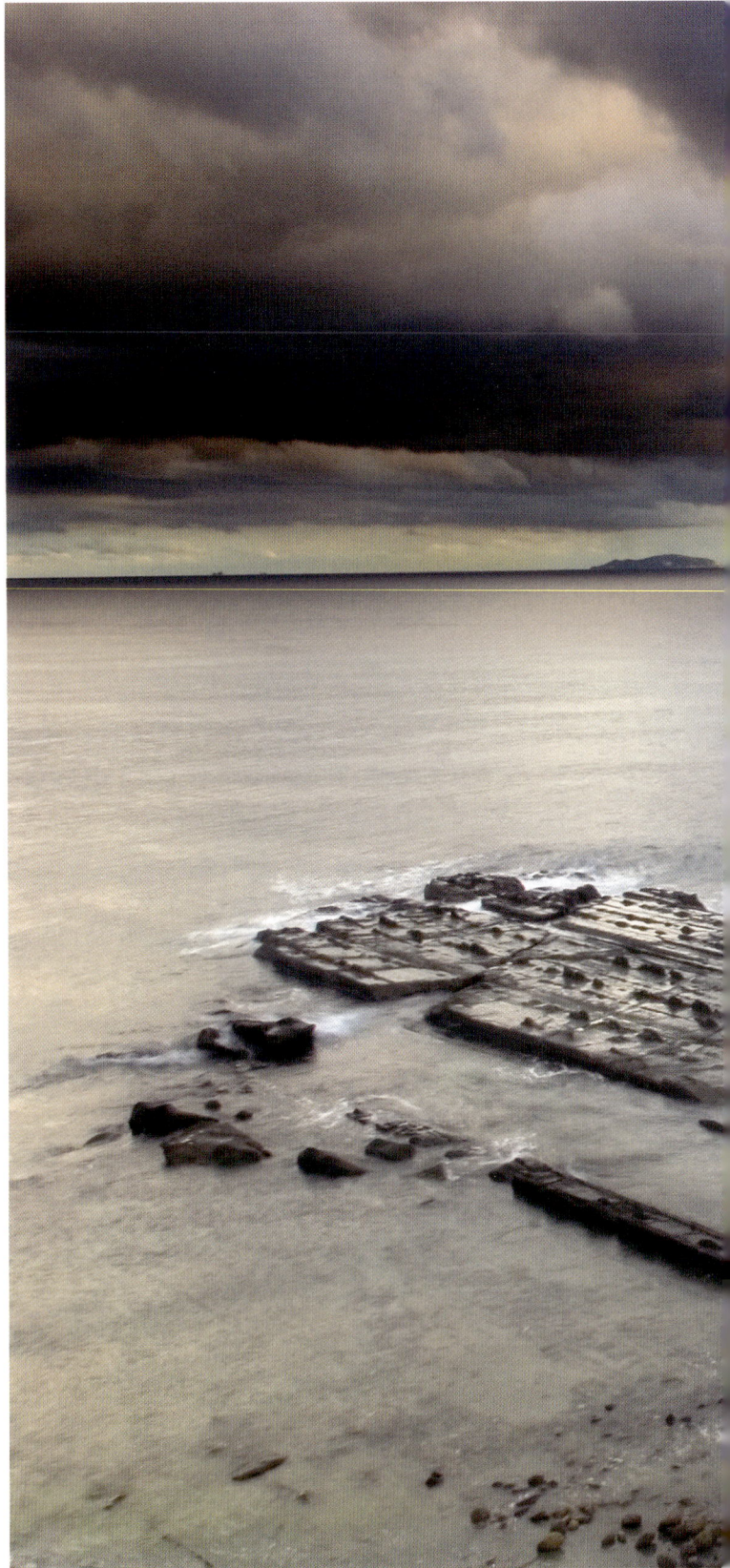

The changing tides in the Campo de Gibraltar uncover examples of Flysch, a rock formation made up of alternating strata of clays and sandstone, eroded unequally.

Perhaps the town where these factors have least impact is Tarifa, because the winds have neither been forced to cover a great distance nor climb summits. Thus, whether the wind is *levante* or *poniente* in summer you do not feel the heat. The people of Tarifa say they don't like Algeciras because it is too warm, although compared to other Andalusian cities it is very cool.

Temperatures

As a reference we can say that in the Campo de Gibraltar the mean annual temperature inland is 15ºC, increasing to 18ºC on the coast, mainly because of the proximity of the two seas, the low altitude and the latitude. In the hottest months temperatures rarely reach the averages of 25ºC on the coast and 29ºC inland. The lowest mean annual temperature is above 10ºC and frost is very exceptional.

Hydrology

Given the proximity of the *sierras* to the coast there are few real rivers to speak of, but streams which dry up in summer, leaving just pools and ponds used by a somewhat irregular but important fauna. To the east the Getares cove is fed by the river Pícaro, and the streams named Marchenilla and La Loba, the first two being in a poor state as they are near to built-up areas. The Morisca stream runs into the sea opposite the island of Las Palomas and the Laurel runs into the Tolmo cove; following on from this we have the Maraber stream and then the river Guadalmesí, the most important watercourse

The Greater Flamingo is a frequent visitor to the marshes near the Strait, which lies on its migratory route.

The lighthouse on the pier at Sotogrande marina, with the Rock of Gibraltar silhouetted in the background.

in the area and which rises in the El Juncal plains. The catchment area on the Mediterranean side of the park ends with the streams called Viña, and then Arroyo, which is inside the town of Tarifa.

The Atlantic side with its valleys and plains is watered by the rivers Jara and Vega and the Salado stream which run into the sea at Los Lances beach. The river Valle meets the sea at Valdevaqueros cove and finally at Bolonia we have the streams called Churriana, in a lamentably polluted state because of the chaotic urban development of the surrounding area, Alpariate and Pulido.

Flora in the Campo de Gibraltar

Two seas and two continents converge at the Campo de Gibraltar. For this reason alone it would be enough to suspect that both foreign and autochthonous species of flora coexist. All this botanical wealth belongs to the chorological *province* Gaditano-Onubo-Algarviense, *section* Cádiz, *area* of Algeciras, with endemic species such as the furze *Ulex borgiae* and the daisy *Bellis rotundifolia*. Many illustrious botanists have come to this area, including Celestino Mutis, Marino Lagasca, Pi Font Quer, Luis Fernández-Galiano and Betty Molesworth, who, until her recent death, lived in Los Barrios, where a

Opposite

The spectacular migration of soaring birds over the Natural Park of the Strait, an annual return journey.

A gathering storm over Los Lances
beach, seen from San Bartolomé.

The White-rumped Swift, together with the Little Swift, is the Natural Park's emblem. They occupy nests built by Red-rumped Swallows in the bunkers dotted along the coast.

park is named after her and she was given the freedom of the town.

About 60 million years ago, at the beginning of the Tertiary period, in the south of Europe and north of Africa there was much leafy woodland with an exuberant wealth of flora. Remaining from this we can still find the skeleton fork fern *Psilotum nudum*, spring bouquet (*Viburnum tinus*), and the spurge laurel (*Daphne laureola*). Towards the middle of the tertiary period the area became drier and woodland was reduced to mountainous and shaded areas with the appearance of new thermophyllous species from Africa such as the dwarf fan palm (*Chamaerops humilis*). In the Pliocene period – from 5 million to 2.5 million years ago – the gap between the two continents opened and closed several times and that was when there was an exchange of flora and fauna between them. Examples of this are the red-berried mistletoe (*Viscum cruciatum*), the narcissus (*Narcissus viridiflorus*), spurge (*Thymelaea villosa*) and the candytuft (*Iberis gibraltarica*), among others. The chameleon (*Chamaeleo chamaeleon*) may have arrived over this natural bridge, although there is a lot of doubt, as may the Barbary partridge (*Alectoris barbara*), the genet (*Genetta genetta*) and the ichneumon (*Herpestes ichneumon*), which could have been brought by the Arabs or Phoenicians. The successive glaciations of the Quaternary pe-

Guadalmesí Tower with battlements, at the mouth of the river of the same name. One of a series of watchtowers spaced along the coast of the Strait.

"La Pantalla" or Screen was built by the military during the Second World War. It was revamped recently and its new image is appreciated by visitors.

riod began some 2.2 million years ago and ended 10,000 years ago, when advancing or receding icecaps from north and central Europe forced some plant species to move, but when the ice moved back they remained isolated in the south. That was the case of the Iberian rosebay (*Rhododendrun ponticum subsp. baeticum*). The maximum exponent of these floral islands is the massif of Sierra Nevada in which a relict flora has found shelter on the summits where conditions of altitude are similar to those of latitude. We

must also pay attention to macaronesic species like the ferns *Vandenboschia speciosa* and *Culcita macrocarpa*, or the tree heather *Erica arborea*; lauraceous plants like the bay tree (*Laurus nobilis*) and the shrub *Viburnum tinus*, also the Saharo-Indian sea rush (*Juncus maritimus*) and the liliaceae (*Asphodelus fistulosus*).

With nearly 1400 taxa, the flower catalogue in the Campo de Gibraltar contains 42 species of fern and 15 orchids, to name the two extremes of plant taxonomy. The endemic Ibero-North

Opposite

Eye of the Moorish wall Gecko, a nocturnal reptile commonly found on rocks and buildings in the Natural Park of the Strait of Gibraltar.

African species make up 12%, the macaronesic 0.4%, local endemic species 5%, Mediterranean flora 23% and European just 1%. Also we must not forget those brought by man from other latitudes and continents and which now belong to our flora. From America came the jimson weed (*Datura stramonium*), the *Agave americana*, the milkweed *Asclepias curassavica,* the relative to the tobacco plant *Nicotiana glauca* and several prickly pears of the *Opuntia genus*. From South Africa came the Bermuda buttercup (*Oxalis pes-caprae*) and the flowering hottentot fig *Carpobrutus edulis* and Cape weed *Arctotheca calendula*, among others. Two endemisms from the Algeciras area also stand out: the *Cytisus tribracteolatus* and a subspecies of the alder (*Frangula alnus subsp baetica*).

The flora in the Park

The dunes, beaches and wetlands form a rather unstable system, at the mercy of the prevailing winds and heavy seas, where the vegetation has to put up with not only the erosive power of these winds but also the salinity, lack or excess of water and the fluctuation of the substratum. In this uncertain environment quite a number of plants have settled, giving the land a chance to stabilize itself. One of these is marram grass (*Ammophilia arenaria*). In this sandy soil you also find *Retama monosperma, Malcolmia litorea, Lotus creticus, pancratium maritimum, Orobanche sp.* In the wetlands we find cord grass *(Spartina densiflora)*, and in floodable areas the sea rush (*Juncus maritimus*) and the less common *Halimione portulacoides, Limonium virgatum* and *Limonium algarvense.*

In the vicinity of the Punta Carnero lighthouse and at the mouth of the river Guadalmesí as well as in the cracks in the cliffs, sometimes resisting sea spray, we find *Limonium emarginatum,* a scarce plant on this stretch of coast, but by contrast quite abundant in Gibraltar. *Asteriscus maritimus* is very near to the sea too and sometimes you can see carpets of them like the one around the western bunker at Getares cove. On the upper parts of cliffs and in gentle hills and meadows a common sight are the dwarf fan palm (*Chamaerops humilis*), purple Jerusalem sage (*Phlomis purpurea*), asparagus (*Asparagus sp.*), and mastic tree (*Pistacia lentiscus*), which, together with *Genista sp.,* form wonderful coastal mosaics. Although it is more usually found in meadows further inland, the French honeysuckle (*Hedysarum coronarium*) is also present. The hairy thorny broom (*Calicotosa villosa*), which crossed over the Strait from Africa, is typical of clayey soils, as is the green narcissus (*Narcissus viridiflorus*). In some very wet meadows we have also found some species of orchids, such as *Orchis laxiflora*, and the beautiful Portuguese squill (*Scilla peruviana*). This particular area has long been subjected to man's influence, especially cattle farming.

In the sierras there is a host of olive trees (*Olea europea*), different types of oak (*Quercus sp.*) as well as eucalyptus and pine plantations (examples of the many ecological crimes committed in this area), which were introduced to replace the cork oak (*Quercus suber*). On poorer stony ground there are flowering heaths (*Erica australis, E.umbellate* and *E.scoparia*) and the Ibero-African endemism - recently discovered in the Cabañeros National Park – the Portuguese dewy pine (*Drosophyllum lusitanicum*), locally called star plant and not flycatcher, as some authors would call it, alluding to other carnivorous plants of the family to which it belongs. To add to these we still have those relics from the past which have survived through the ages such as the Phoenician juniper (*Juniperus phoenicea*) and juniper (*Juniperus oxycedrus subsp macrocarpa*) around Punta Camarinal.

Nor should we forget the flora growing next to the rivers, especially the largest one, the Guadalmesí, where there are alder (*Alnus glutinosa*), white willow (*Salix alba*) and white poplar (*Populus alba*), together with oleander (*Nerium oleander*) and a few cork and wild oaks. Towards the end, almost at the river mouth there are some giant reed or cane groves (*Arundo donax*).

The fauna in the Park

The variety of animal species in a biotope depends, among others, on its altitude, latitude and plant diversity. In this park rocky or wooded areas are few, but beaches, rivers, cliffs, marshes meadows and scrubland are plentiful. Animal life, therefore, must adapt to these conditions accordingly. In this section we are going to deal exclusively with land and freshwater fauna and will only refer to marine life in exceptional cases. Five species stand out among the multitude of fauna: the otter, the White-rumped Swift, the Little Swift, the fish *Aphanius baeticus* and the monarch butterfly, which we will come back to later on.

On the sandy beaches the area left by the outgoing tide is plied by a jumping crustacean known locally as the sand hopper (*Talitrus saltator*) which feeds on residues which the tide brings in. Years ago, when stormy seas used to leave huge quantities of algae behind, the beach looked to be boiling. The sand hoppers invaded them in their thousands and they in turn were eaten by wintering or migrating shore birds. Among the considerable number of different vertebrates we can find the spurge hawkmoth caterpillar (*Hyles euphorbiae*) feeding on the spurge plant (*Euphorbia paralias*) growing on

the dunes. The latter are the favourite biotope for some reptiles such as lizards and the western spadefoot toad (*Pelobates cultripes*).

On beaches and in marshes some birds can be observed clearly all the year round. Among these local residents are the Kentish Plover (*Charadrius alexandrinus*), Cattle Egret (*Bubulcus ibis*), Little Egret (*Egretta garzetta*) and the Grey Heron (*Ardea cinerea*). Summer visitors include the White Stork (*Ciconia ciconia*) and winter ones are the Sanderling (*Calidris alba*), and the Caspian Tern (*Sterna caspia*). Migrant birds such as the Little Tern (*Sterna albifrons*), Audouin's Gull (*Larus audouini*), Greater Flamingo (*Phoenicopterus rubber*) and Osprey (*Pandion haliaetus*) also drop by. Very rarely you can also spot the Razorbill (*Alca torda*), Puffin (*Fratercula arctica*) and the Storm Petrel (*Hydrobates pelagicus*).

The avifauna on the shingle beaches, reefs and islets is also of interest. Most are migrants or winterers: shore birds like the Curlew (*Numenius arquata*), Turnstone (*Arenaria interpres*), Ringed Plover (*Charadrius hiaticula*) and Common Sandpiper (*Actitis hypoleucos*) and also Audouin's Gull (*Larus audouini*). Another migrating gull is the Yellow-legged Gull (*Larus michaellis*), and there are winterers like the Black-headed Gull (*Larus ridibundus*), Mediterranean Gull (*Larus melanocephalus*), Lesser Black-headed Gull (*Larus fuscus*) and the Herring Gull (*Larus argentatus*), which until quite recently was thought to be the same species as the Yellow-legged Gull.

On the cliff at Punta Carnero a pair of Black Kites used to nest in the fifties and sixties, while in the seventies they were replaced by a pair of Ravens. Meanwhile, these have lost their place to two Kestrels (*Falco tinnunculus*), which nest there now. During the month of May and early

Overleaf

The Eyed or Jewelled Lizard is now being preyed upon, due to the scarcity of rabbits.

The Red-berried Mistletoe (*Viscum cruciatum),* also found on the Moroccan shoreline, grows on limestone soils.

June when the migrating soaring birds are still arriving at the coast in the park, these little falcons defend their cliff very aggressively and fight off any Short-toed or Booted Eagle, kite or even Griffon Vulture which dares to approach their breeding area. Some of these raptors arrive at the coast absolutely exhausted and have to make a supreme effort to escape from sustained attacks. Kestrels also nest in some of the beacon towers along the Strait, hunting among migrating passerines to feed their young. Another, nocturnal, raptor, living on the cliff is the Little Owl (*Athene noctua*), which nests among the rocks and in abandoned farm buildings. On the clay tracks at dusk you may also spot the Red-necked Nightjar (*Caprimulgus ruficollis*) which breeds in the park.

Given the paucity of crags and crevices in which to build nests, many birds use buildings such as the beacon towers. The military bunkers also attract many species, one of the most sig-nificant being the White-rumped Swift (*Apus caffer*). This bird actively evicts the Red-rumped Swallow (*Hirundo daurica*) from its nest and occupies it, ejecting eggs and even chicks (Barrios, F., 1994). It is without doubt the most famous bird in the Natural Park of the Strait and the surrounding area. They need protecting by taking steps to stop people going into the bunkers – known locally as machine gun nests – so that they can breed in peace. These constructions are also home to Red-rumped Swallows, Wrens (*Troglodytes troglodytes*) and Kestrels (*Falco tinnunculus*). Another really fascinating animal, and perhaps the least known, is the bat which has found in these bunkers an ideal substitute for caves. As they have tremendously thick walls and most are underground with only small slits opening to the outside world, the temperature inside the bunker remains practically constant throughout the year – one of the essential conditions for the survival of these little known animals. Indeed, some of these machine gun bunkers are vital as wintering shelters for some unique species of bat in Spain. In other buildings you can find nesting Swallows (*Hirundo rustica*), Barn Owls (*Tyto alba*), Little Owls *(Athene noctua)* and many types of reptile which hibernate or hunt there such as the Mediterranean gecko (*Hemidactylus turcicus*) and the Moorish gecko (*Tarentola mauritanica*).

Opposite

The American monarch butterfly has settled in the area, using the Canary Islands as an intercontinental stepping stone. It is possibly the most common of the *Lepidoptera*.

Shot during violent courting of a pair of *Psammodromus algirus*, or Mediterranean lizards, in which the male holds the female firmly.

In the shrub and pasturelands there is an abundance of reptile species, pride of place going to the jewelled or ocellated lizard (*Lacerta lepida*). As a species it has come under threat due to the disappearance of the rabbit, which is similar in size, and has now become prey to eagles, ichneumons and foxes, among others. That is why, from being a prosperous species in the fifties, it is now classed in the *unknown* category in some areas. Although this reptile is not too numerous in the park, we can say that it is certainly one of the few areas where this magnificent species can still be observed. This is due to several factors: the limited number of vehicles using the tracks and the slow speed at which they travel on these very uneven, bumpy surfaces, and also the number of containing walls, originally built next to the tracks to prevent landfalls, but which now due to their age have many cracks in which the reptiles can live. The Montpellier snake (*Malpolon monspessulanus*), the ladder snake (*Elaphe scalaris*) and other smaller ones share this biotope with one of the major predators of the park: the ichneumon (*Herpestes ichneumon*). This animal, together with the fox (*Vulpes vules*) is at the top of the food chain in this biotope. The rabbit (*Oryctolagus cuniculus*) has been almost eliminated by predators and hunters and the current population is at its lowest ever. Another representative of African fauna in the park is the Algerian hedgehog (*Aethechinus algirus*) and is an implacable devourer of snakes.

The strictly nocturnal is one of several
African vertebrates introduced by the
Phoenicians or the Arabs.

A lush springtime meadow, with a myriad of flowers and mount San Bartolomé in the background, quite near the ancient Roman city of Baelo Claudia.

The rivers and their spheres of influence are the most important habitats in the park because they contain three of the most interesting animals (even on a national scale) of its entire fauna, among them an endemic Iberian fish – a toothed carp – (*Aphanius baeticus*), and the monarch butterfly (*Danaus plexippus*), originally from America. In summer almost continuously flowing water courses, such as the river La Vega, are reduced to pools containing species such as *Aphanius baeticus*, which tolerates brackish water. This is a new species among Iberian fauna, differing from the native Mediterranean toothed carp (*Aphanius iberus*). It has only been detected in certain parts in the provinces of Huelva, Sevilla and this area of Cádiz and is classed as an *endangered species*. Its recent discovery (Clavero,M., *et al.*, 2002, *Quercua*) was a wonderful surprise and reaffirms how vital this park is as a key to the survival of

endangered fauna, not only as a refuge for this fish but also for the otter (*Lutra lutra*). Recent studies have shown that the rivers in the area are essential for the intercommunication of local otter populations in the different catchment areas of the Campo de Gibraltar. Apart from these two significant species we can add a third, namely the valuable population of monarch butterflies on the banks of the river Pícaro which runs into the sea at Getares cove. On the banks, quite near the mouth there is an area of scrub with brambles, mastic trees and fan palms and among the flora to be found there is a wide area of *Asclepias curassavica*, a plant of the curare family, originating in South America, and on which the caterpillar of this beautiful butterfly feeds. It is now the most common lepidopteron in the Campo de Gibraltar and in cities like Algeciras it is often seen in streets with little traffic.

Other typical amphibian species to be found in and around the water courses are the painted frog (*Discoglosus jeanneae*), fire salamander (*Salamandra salamandra*), the endemic Andalusian triton (*Pleurodeles waltl*), the common toad (*Bufo bufo*) etc.

On the crags it is the birds which capture our attention. A total of 50 pairs of Griffon Vultures are nesting in the *sierras* of San Bartolomé and La Plata. The road runs right next to La Plata and on a huge sandstone crag you can easily see the vultures feeding their young or just sitting there on the Laja Ranchiles rocks. There is still a nesting pair of Eagle Owls in San Bartolomé and, until a few years ago there was a pair of Egyptian Vultures (*Neophron percnopterus*), Bonelli's Eagles (*Hieraetus fasciatus*) and Peregrine Falcons (*Falco peregrinus*). In the sixties, in the *sierra* of La Plata, we observed an Eleonora's Falcon attacking and repeatedly chasing away a common buzzard. It was at the end of June and, some years later, when we had acquired more knowledge, we wondered if it could have been breeding, because its behaviour pointed towards that.

Among all this interesting bird life two species of swift stand out: the White-rumped (*Apus caffer*) and the Little Swift (*Apus affinis*). The White-rumped Swift is a migrating bird originally from Africa, discovered in 1964 and at first thought to be a Little Swift. As mentioned before it occupies Red-rumped Swallows' nests by force so that these have changed their nesting habits from rock cavities to buildings and at the moment we know of no Red-rumped Swallow's nest being occupied in rocks. However, the White-rumped Swift also breeds in these buildings, which in the park are the abandoned bunkers.

The Little Swift is also a migrant from the African continent, observed for the first time in 1983 and recorded as a breeding bird in the *sierra* of La Plata in 2000. The two swifts both have white upper rumps, but the Little Swift is smaller and has a square-cut tail, not forked like the White-rumped Swift's.

Marine fauna would not normally come under the fauna of the park, but I should like to make an exception with the grouper (*Epinephelus guaza*). As I said before, I used to go underwater fishing in these amazing waters of the Strait from the end of the fifties until 1974. At first I would see groupers at a depth of about two metres and in the crevices where they took refuge I could often see two or three together. But over-fishing and pollution have made the species quite rare and the few that are left are seen 25 to 30 metres down. Only near Tarifa island where there is a marine reserve can you see magnificent specimens at a shallower depth. We have heard that in the south of France grouper fishing was banned for a few years and the population has recovered, especially with the creation of artificial reefs which prevent over-fishing. It would be a good idea to copy this way of protecting and saving this fine animal from extinction in the waters of the park. It would certainly be a good advert for tourists who could then come to these rough waters to observe these splendid fish.

I am quite aware that a lot more could be written about the Natural Park of the Strait, but let us leave it to other authors to write a long monographic work on these 9,880 hectares which contain such exceptional flora, fauna and geology. Certainly the 9,257 hectares of sea will produce some endemism waiting to be discovered, although just to enjoy the splendour of its depths it is worth putting on your goggles and fins to drift in the cold currents down to a hidden paradise, where some camouflaged sole will watch the silhouette of a friend glide by.

Perhaps the most significant aspect of the Park has not, so far, been mentioned: migration. Because of its geographical situation the Strait of Gibraltar is one of the most important enclaves in the world for the migrants which pass through it – fish (tuna, mackerel, bonito etc), reptiles (turtles), birds (sea and land species) mammals (whales, killer whales, dolphins etc) and, surely a subject for future study, the invertebrates (butterflies and dragonflies). In this book we are going to deal with migrating soaring birds but books could be written about the migration of other animal species. Thanks to migrating cetacean species, some prosperous businesses, specializing in whale and dolphin watching in the Strait, have sprung up in the towns of Algeciras, Barbate, Gibraltar, La Línea de la Concepción and Tarifa, giving the tourist an additional reason to come here.

In the following chapters we shall put all our efforts and knowledge into describing the migration of soaring birds in great detail, so that readers will be able to appreciate the full importance of this very special corner of the world.

The constant, strong *levante* wind
overwhelms the vegetation in the area
and makes it lean towards the west.

Nothing surprises one more than the brilliant flora in May and June: it is like a greenhouse gone mad: flowers of all colours, like perfumed cups of rubies, amethysts and topazes full of solar light, tempting the stranger at every step. They bloom and blush without the native noticing them.

Richard Ford
Handbook for travellers in Spain and readers at home, 1845

Los Alcornocales: a resting place

Geographical limits and relief

In this chapter, dedicated to Los Alcornocales Natural Park, topics such as the climate and meteorology will not be touched on as they have already been described in the previous chapter on the same area.

Los Alcornocales Natural Park is made up of a series of *sierras* grouped together at the eastern end of the province of Cádiz, running from north to south between the depressions of the river Barbate in the west (the area of La Janda and Medina Sidonia) and the rivers Palmones and Guadarranque in the east; the mountain passes of Gáliz, Palomas and Algarrobo are the northern limit and in the south it is bounded by the recently created Natural Park of the Strait. The *sierras* are part of the south western end of the Betica mountain range, corresponding to the Alpine orogeny. Of its total surface area of 170,025 hectares, nearly 75% is privately owned land made up of large estates given over mainly to agriculture, hunting and the cork industry.

The Park is situated within the following municipalities of the province of Cádiz: Alcalá de los Gazules, Algar, Algeciras, Castellar de la Frontera, Jerez de la Frontera, Jimena de la Frontera, Los Barrios, Medina Sidonia, San Roque, Tahivilla and Tarifa. A small piece of the province of Málaga, belonging to the township of Cortes de la Frontera, also forms part of the Park.

The land, as a whole, is made up of parcels of sandstone, of allochthonous origin, "floating" on marls and clays like ice cubes in water. This predominance of sandstones and clays forms the basis of a landscape of moderate altitudes (1,120 m at El Picacho), deep valleys and both varied and exuberant flora. The soil's main character-

Algar

**Cortes
de la Frontera**

**Alcalá
de los Gazules**

Medina Sidonia

**Jimena
de la Frontera**

**LOS ALCORNOCALES
NATURAL PARK**

**Castellar
de la Frontera**

San Roque

Los Barrios

Tahivilla

Algeciras

Atlantic
Ocean

Tarifa

Mediterranean
Sea

Ojén, Luna, Bujeo and finally, sloping gently down to the sea, the Sierra de Fates.

As with the Natural Park of the Strait, the latitude, proximity of two seas, the siliceous soils and the joining of two continents give Los Alcornocales and, indeed, the whole region some exceptional characteristics as regards climate, fauna and flora. Furthermore, the general state of conservation is good, thanks to a profitable cork industry, which is almost exclusively extraction as most transformation of the cork takes place outside the area.

Vegetation and flora

The concurrence of a number of favourable elements makes for an extraordinary variety of flora whose origins are quite different both in time and in geographical location. There are still remnants from the Tertiary period, about 60 million years ago, such as *Laurus nobilis, Daphne laureola, Diplazium caudatum* and *Viburnum tinus*; others arrived via the last glaciation from central Asia like the rhododendron and *Ruscus hypophyllum*; from the Macaronesian islands (Azores, Canaries, Cape Verde and Madeira) came *Vandemboschia especiosa* and *Davalia canariensis* and finally, from the Atlantic region, *Blecnum spicant*.

It would be rather tedious to list the emblematic species of the Park, there being so many, but if one had to choose the ten which, although without any scientific criteria, best illustrate the maritime and African vocation of these southern lands, they would be: *Psilotum nudum, Chistella dentate, Laurus nobilis, Quercus suber, Cytisus gaditanus, Arisarum proboscideum, Drosophyllum lusitanicum, Rhododendron ponticum subsp baetica, Ilex aquifolium* and *Scila peruviana*.

At the head of the streams which are locally referred to as *canutos* or canes, alluding to

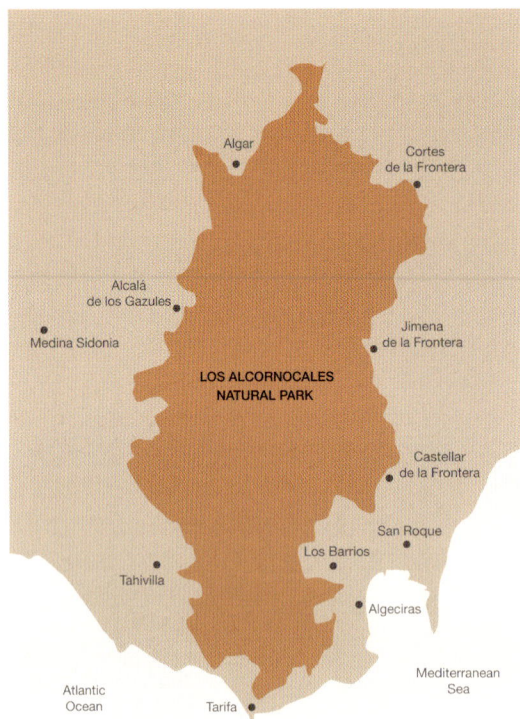

istic is its acidity which is the key to the abundance of cork oak forests covering most of the *sierras*.

This Natural Park is notable for its complicated relief because, although altitudes are relatively low, the distance from the coast is short and an altitude of 800 m can be reached after only 8 kilometres. This collection of low altitudes with deep valleys can make the visitor feel as if they are in an alpine landscape with no snow but with an overwhelming amount of dense vegetation. In a line from north to south we have the Sierra del Aljibe (1,092 m), Momia, Montecoche, Blanquilla, Junquillo, Sierra del Niño, Sequilla,

Opposite

Diplazium caudatum hidden away in a "canuto" or stream with other ancient *pteridoflora*. In danger due to climate change.

the vision you would have through a cane which grows there, the vegetation forms a kind of tunnel. The sun's rays pass through two filters before acting on the flora on the ground. The first filter is through the tops of alder (*Alnus glutinosa*) and Algerian oak (*Quercus canariensis*) trees; a second filter is provided by rhododendron (*Rhododendron ponticum subsp beticum*), laurustinus (*Viburnum tinus*), alder buckthorn (*Frangula alnus subsp baetica*) etc. This canopy keeps the *canuto* in permanent shade, causing a thermal insulating effect so that the temperature varies by only a few degrees, both in winter and in summer: the average near the coast is about 20º C all year and about 18º C further inland. In addition the thermo regulating effect increases the relative humidity to over 90%, due also in part to the mists mentioned in chapter I.

In the vegetation of Los Alcornocales it is the pteridophytes which are the most spectacular plants and mark the difference. The rich variety consists of forty-two species, including tropical ones like *Chistella dentata* which is found in Crete, two places in the Iberian peninsula and in three of the Macaronesian archipelagos (Salvo, E. 1990 Guía de helechos). Another species which will amaze the botanist is *Psilotum nudum,* discovered in the *sierras* of Algeciras in the 60's by the distinguished botanist Betty Molesworth. This fern, rupiculous in this area, is rarely found outside the zone between the tropics of Cancer and Capricorn and is the only representative in the Palaearctic region, the nearest population being on the Cape Verde isles, 4,000 km away. As with the two previous examples, *Diplazium caudatum* also came to light, thanks to Betty Molesworth, in 1966. This pteridophyte is a Macaronesian endemic which in continental Europe is only found in

Psilotum nudum a fern relict found in the early 1960's by Betty Molesworth, whose botanical discoveries reaffirmed the importance of pteridophytes in the Campo de Gibraltar.

isolated parts of Los Alcornocales. It requires strict conditions of humidity, shade and very slight temperature changes. *Culcita macrocarpa* is another Macaronesian endemic present in Los Alcornocales, and also in Galicia and Asturias, but is under threat in the Park because of the former use of the fluff covering the rhizomes as an anti-haemorrhage agent. There is a *canuto* in the Dehesa de Ojén in Tarifa where, walking a short distance in a straight line we can observe, among other ferns, *Diplazium caudatum, Vandesboschia especiosa, Pteris incompleta*

Opposite

There are a total of 27 different species of orchids in The Alcornocales Park, mainly on limestone soils. Here we see a Woodcock Orchid.

Constant humidity, due to ever-present mists, and temperatures with little seasonal variation, have helped to preserve flora relicts from the Tertiary period.

The Cork Oak tree occupies most of the forests of The Alcornocales Park.

and *Culcita macrocarpa*. It may well be the most important area in the Palaearctic region from the pteridological point of view.

These sandstone *sierras* are mainly covered by extensive cork tree (*Quercus suber)* forests and other oak varieties such as *Quercus canariensis, Quercus faginea* and the holly oak (*Ilex aquifolium*).

The cork oak tree, a typical western Mediterranean species growing in sandy soils, and known locally as *chaparro*, has survived to the present day, forming dense woodlands thanks to the profitable cork extraction industry and the poor agricultural quality of the soil. It shuns the cold and needs average temperatures of over 14ºC, minimum winter temperatures above 0ºC and at least 600mm annual rainfall. The Greeks knew it as the "crust tree" and 93% of world cork production comes from Spain, Portugal and Morocco. It is a perfect example of nature's adaptation to the environment, since the cork is nothing more than insulating material to protect the tree from the frequent forest fires which were the scourge of these western Mediterranean lands, especially in the summer months. When a cork oak is stripped it is defenceless against the cold, fire, invertebrates and diseases. So, in order to allay these difficulties cork stripping (*la saca*) takes place from the second half of June on, when the atmosphere has heated up and there is no fear of rain or low temperatures, until the first fortnight in August. In Spain, cork oak forests cover about 360,000 hectares, most of them on large estates. In Los Alcornocales they are situated almost exclusively in the *sierras*.

The extraction of cork or stripping is carried out using the same methods as two centuries ago. It needs considerable hand labour and no machinery intervenes until the cork is transported to the factories. Skilled workers or *corcheros*, under the supervision of the *manigero*, separate strips of cork from the tree with a special axe, so as not to damage the trunk, because any gashes would allow disease and parasites in. These strips are then collected by the *recogeadores* and piled up to be loaded onto mules which are taken by the *arrieros* to clearings, where they are weighed and loaded onto lorries. To quench their thirst brought on by the hard work and fierce summer sun, the *aguaores*, usually the youngest workers, supply the cork strippers with water from earthenware water jars. Groups of workers used to be more numerous but have been cut down to reduce labour costs. About 40 years ago a group was made up of *corcheros*, *rayaores* who, after the stripping process, would make a vertical incision in the trunk to allow for expansion and reduce the unevenness of the bark. The *recortaores* cut up the cork into smaller strips so that the mules can carry it. The *apilaores* pile these up and then the *recogeadores* load them onto the mules which are led away by the *arrieros*. Let us not forget the

aguaores and also a cook for the whole group! Nowadays, however, there are no longer any *rayaores* and the work of the *recogeor* and *apilaor* is carried out by one man.

Cork is first stripped from a cork oak after 20 years, depending on the state of the tree, when it has a diameter of about 20cm. This first stripping or *bornizo* is not suitable for the cork stopper industry but will be ground up into cork agglomerate. After that the tree is stripped every nine years. Thanks to the cork stopper industry which absorbs 90 % of the production, cork extraction continues to be viable, but the introduction of plastic and rubber stoppers is worrying both for the industry and the conservation of these forests.

Apart from cork the cork oak was a raw material source for charcoal and also what is called *la casa* or *madre*. This is the layer of tissue which later develops into wood or cork. It is tannin–rich and was used in the tanning industry in the area. In the past this gave rise to quite considerable deforestation.

One very worrying aspect for the Park these days is the state of health of the forests, which, if no drastic action is taken, could lead to their disappearance within the next 30 years. Several experts agree with this figure, yet it does not seem to have caused estate owners, trade unions, official organisms or ecologist groups to become unduly worried.

1987 saw the publication of an *Edict about how to proceed in wooded areas affected by the Mediterranean Hypoxilon*, a fungus which was blamed for each and every disease of the cork oak forests, although it was soon evident that more than a disease it was a symptom of bad forestry management due to the absence of disease control, poor cork stripping, droughts, etc. Soon

The loads of cork strips are taken by mules to clearings accessible by lorries for transport to the factories.

A muleteer readies the load of cork to be carried by mules through the dense woodland of cork trees growing on steep slopes.

another pathogenic fungus, *Diplodoa mutila* appeared and later on *Ceratocystis fagacearum.* Traditionally, blame for the deterioration of the cork oak forests has also been attached to the gypsy moth (*Lymantria dispar*), a lepidopteran whose insatiable caterpillars eat the leaves off the trees while in turn they are preyed on by a coleopteran called *Calosoma psicophanta.* Several studies have given rise to the suspicion that these pathogenic agents are only taking advantage of the weakness of the forests and the real culprit is, as we mentioned before, poor management leading to ageing of the trees themselves.

Forest fires are beneficial to the forests but bad for their owners according to José María Sánchez, a Park technician, speaking at a conference on the Flora and Fauna of the Campo de Gibraltar. Up until the 1960's the prophylactic role of forest fires was carried out by the charcoal burners. Weak and diseased trees were taken out, cut up and made into charcoal on the spot, thereby preventing the spread of disease and keeping the forest healthy. When the charcoal burners went for good, the sick trees, attacked by fungi, stayed where they were and their spores easily propagated disease among other trees. But let us not delude ourselves: these evils are due to the very poor or complete absence of forestry management. Every nine years the owners of these vast cork oak estates have simply invested the profits elsewhere or just prepared the ground for the next cork stripping. Too much animal farming prevents regeneration through acorns and over the last few decades shoots from the tree stumps or roots have had too much of an advantage over acorns, which end up being eaten by herbivores. To the untrained eye the cork oak forests look healthy but nearly all the trees originate from stumps or roots and, although at first sight they look young, they are, clone-like, in fact as old as the most decrepit trees in the wood. The cork oak forests are on Death Row, they are old and defenceless in the face of attacks by fungi and invertebrates, unscrupulous axes and the proliferation of estates where deer and cattle do away with the oval-shaped hopes of salvation, the acorns. It is what is known locally as *la seca*, a terrible word which perfectly sums up the medium term future of Los Alcornocales. Some people envision an even more pessimistic scenario: if, as some experts forecast, rubber and plastic stoppers gain more ground and working the cork oak forests becomes less profitable, then this demoralizing future has already arrived.

The larvae of the *Calosoma psicophanta* devours
the caterpillars of the moth known locally
as the hairy lizard, and has devastating effects
on the development of the Cork Oak tree.

The Rhododendron flowers twice yearly in Los Alcornocales, although the december flowering is less well-known.

In the undergrowth on the sunny side of the cork oak forests there is hairy thorny broom (*Calicotome villosa*), *Ulex borgiae*, sage-leaved rockrose (*Cistus salvifolius*), Jerusalem sage (*Phlomis purpurea*) and on the shady side and near streams we find heather, broom and common ferns (*Pteridium aquilinum*).

Together with the cork oak, although in the damper and shadier areas, there are Algerian oaks (*Quercus canariensis*) which also prefer sandy soils. This tree has marcescent leaves, i.e. the dead leaves stay on the tree until replaced by new ones in spring. These trees were lopped for wood by the charcoal burners and with some of their main branches missing they have a rather unusual shape. They need more water than cork oaks and are found in damper areas near streams and on the northern side of the *sierras*. They are more prone than cork oaks to supporting epiphytes such as *Davalia canariensis, Polipodium sp*, pennywort (*Umbiliculus rupestris*) etc, as their bark is not taken off every nine years like the cork oak's. Both cork oak and Algerian oak forests are home to a host of laurustinus, phillyrea (*Phillyrea latfolia*), butcher's broom (*Ruscus aculeatus*), and wild madder (*Rubia peregrine subsp longifolia*). Creepers and lianas such as *Clematis cirrhosa, Lonicera hispanica, Aristolochia baetica, Aristolocia longa*, ivy (*Edera helix*), strawberry trees (*Arbutus unedo*) and laurestinus are also commonly found in Algerian oak woods.

A *canuto* in the Sierra de Algeciras,
containing remnants from the Tertiary
period such as bay laurel, laurustinus,
strawberry tree, etc…

The other oak present, the Portuguese oak (*Quercus faginea*) is more sandstone-shy than the Algerian oak but also prefers shade and damp conditions for growth. It is accompanied by the mastic tree (*Pistacia lentiscus*), myrtle (*Myrtus communis*), hawthorn (*Crataegus monogina*), wild pear *(Pyrus bourgaeana)* etc

The holm oak appears in calcareous soils in fields in the northwest part of the Park and, compared to the cork oak, does not form extensive masses of forest. Its presence is more incidental and not directly related to the Aljibe sandstone soils where the undergrowth is comprised of carobs (*Ceratonia siliqua*), mastics, wild asparagus and hawthorns etc.

The *acebuche* or wild olive (*Olea europaea var. silvestris*) populates the poorly drained clay soils which get saturated with heavy rain and crack up when the water evaporates in summer. It is a vital tree for the autumn migrant birds because its fruit provides substantial supplies of fat. The same goes for butcher's broom and the mastic, strawberry, holly and hawthorn trees. When they are very old their trunks provide hollows for the Little Owl, Tawny Owl and sometimes genets. Asphodels (*Asphodelus sp*), Spanish oyster plants (*Scolymus hispanicus*) and wild asparagus (*Asparagus albus*) grow near the wild olive.

The scrub is made up of woody plants which crop up after destruction by forest fires. It consists of quite extensive masses of heather, cistus, mastic trees, broom, dwarf palms on poor soils. As with the wild olive groves these areas are extremely important for resident and migrating birds, especially *Passeriformes*. It is the biotope for the largest Iberian carnivorous plant, the Portuguese dewy pine (*Drosophyllum lusitanicum*) whose name alludes to its liking for flies. On the bare summits of the *sierras*, lashed constantly by strong winds, conditions limit the vegetation to *Quercus lusitanica*, heather (*Calluna vulgaris*), heath (*Erica umbellate*), poplar-leaved cistus (*Cistus populifolius subsp major*) and *Halimium alyssoides*.

The pastureland is in downgraded areas which have expanded thanks to livestock and agriculture, yet, on the other hand, its vegetation is one of the richest and most diverse in temperate zones. In an area of one square metre you can come across up to 40 different species (Blanco, R. *et al*, 1989 Sierras del Aljibe y del Campo de Gibraltar): *Scilla peruviana, Iris sp*, dwarf morning glory (*Convolvulus tricolour*), gladioli (*Gladiolus iliricus*), paper white narcissus (*Narcisus papyraceus*), Spanish eyes (*Erodium sp*), Bermuda buttercup (*Oxalis pes-caprae*), French honeysuckle (*Edysarum coronarium*), to name but a few.

The hydrology of the Park has one important characteristic: it has two catchment areas, the Atlantic side and the Mediterranean side. Rivers which flow into the Mediterranean are the Miel, Guadalmesí, Palmones, Guadarranque and Hozgarganta, while the Jara and Barbate flow into the Atlantic. As the *sierras* are not far from the sea, these permanent water courses are not long and come down from medium altitudes in a short distance.

Because of livestock and agrarian interests the river banks and adjacent areas have been under intense human pressure and the deterioration is quite evident, witness the rivers Palmones, Guadiaro, Guadarranque and Hozgarganta. These rivers are examples of the seasonalness of the area and, depending on the river and the year, water may stop flowing, leaving deep pools where the river fauna finds refuge. Due to human pressure the woods along the river banks

Opposite

The Portuguese Dewy Pine, heath and poplar-leaved cistus are common on poor quality soils.

have deteriorated along the mid and lower sections and what we have are discontinuous clusters of white poplars (*Populus alba*), ash (*Fraxinus sp*), elms (*Ulmus minor*) and willows (*Salix sp*). To meet the needs of the local population and industry, especially in the Bay of Algeciras, a number of reservoirs have been built: Almodóvar, Charco Redondo, Guadarranque, Los Hurones, Barbate, Guadalcacín and Celemín.

Fauna

Just as we have seen with the flora, among the fauna in the Park are species typical of Mediterranean woods as well as others originating in Africa such as the genet (*Geneta geneta*), the mongoose (*Herpestes ichneumon*), the White-rumped Swift (*Apus caffer*), Little Swift (*Apus affinis*), domestic gecko (*Hemidactilus turcicus*), Algerian hedgehog (*Aethechinus algirus*) etc, which have either been introduced – like the genet and mongoose – or have "jumped ship" like the swifts and the gecko. However, we are not only referring to the "jump" across the Strait, but also across the Atlantic Ocean, as is the case of the monarch butterfly, and, perhaps the most spectacular example, Rüppell's Vulture (*Gyps ruppelli*).

The rich variety of species of vegetation gives rise to a legion of herbivores which are preyed upon by reptiles, birds and mammals. At the very top of this food chain are the fox (*Vulpes vulpes*), mongoose, Griffon Vulture (*Gyps fulvus*), Bonelli's Eagle (*Hieratus fasciatus*), Sparrowhawk *(Accipites nisus)* and nocturnal raptors like the Eagle Owl (*Bubo bubo*).

The clay soils and pastureland should in theory be the perfect ecosystem for the rabbit (*Oryctulagus cuniculus*) but the sad reality is that it is rarely seen in the Park, although outside it there is a reasonable quantity. There are areas,

The Fox is one of several opportunist predators, abundant in all biotopes in spite of being under constant pressure, as it is always in the hunter's sights.

The Ichneumon is the only mongoose, of African origin, with strictly daytime habits.

like the former lagoon of La Janda, right next to the Park, where there is an abundance of rabbits, so much so that the Spanish Imperial Eagle is being reintroduced into the area from there. Myxomatosis and later pneumonia have done away with 85% of the population, leading to the decline of many predators. Some of these enter into the sad category of being the most endangered feline species in the world, i.e. the Spanish lynx, and the most endangered bird of prey in Europe, i.e. the Spanish Imperial Eagle.

Hares (*Lepus capensis*) have also seen their population diminish because of human activity and the lack of rabbits. However, there are game reserves on estates where they are still numerous because they represent an important source of income for the owners, together with the Red-legged Partridge (*Alectoris rufa*), which just about manages to survive, despite the pressure this typical Iberian species is put under from hunters. They are rarely seen in pastureland although in recent years we have seen some taking refuge in scrubland and woods.

Nevertheless, there is one bird, the Cattle Egret (*Bubulcus ibis*), which has successfully adapted to pastureland, thanks to the presence of a large number of livestock and the sown fields. There was even a breeding colony near the area, on the cliffs at Barbate, which it shared with Little Egrets and gulls. After colonizing southern Europe, this African heron crossed over the Atlantic and is now a common sight in South America, for example in Bolivia, as is the common sparrow.

Another conspicuous inhabitant of the meadows is the Bee-eater (*Merops apiaster*) with its colourful African plumage, although it is European by birth. This migrant bird arrives towards the end of March and excavates its nest in sandy outcrops or, failing these, in the clayey soils. Other, similar-sized birds, such as the European Roller (*Coracias garrulous*) and the Hoopoe (*Upupa epops*), are rarely spotted in Los Alcornocales and usually only in passage.

In the scrub and forest there is one large mammal which stands head and shoulders above others: the red deer (*Cervus elaphus*). It was re-introduced in the 1950's and 60's into many estates to make them more profitable. This only required fencing and a minimum of vigilance to avoid poaching, quite rife in the area. Nowadays, and especially since Spain joined the EU, this animal has become an important source of income for many estates, whose size had made them economically unviable. Hunters from all over Europe arrive in the area at weekends, kill the game and are back home on Sunday evening. Hunting as an industry has contributed to genetic deterioration in the species: the strongest and most vigorous animals are shot; there is an excessive number of deer per hectare, leading to over exploitation of the vegetation, thus harming the cork oak forests; the roe deer (*Capreolus capreolus*) population has suffered because of competition for the same food, and disease has spread. In the early 70's on some estates, such as La Almoraima, deer were brought in from central Europe to "improve" the species, but the real ulterior motive was to increase the size of the game to make it more attractive to hunters who were prepared to pay top prices for their trophies. As luck would have it, poachers prevented these "beneficial" genes from being passed on to new generations of deer and they were the first to disappear, victims of greed inspired by their considerable size.

The roe deer is the Park's star attraction. Its population in these forests is the southernmost in Europe and because of geographical isolation (the nearest roe deer are in Sierra Morena) it has some special characteristics. Although its genetic status does not confer upon it the category of subspecies, it is representative of an ecotype belonging to the *sierras* of Cádiz and Málaga. It is smaller, grey in colour all the year round and does not have the typical white mark on its throat. Both in the rutting and antler shedding seasons and when they give birth, they do not coincide with the rest of Iberian roe deer, possibly because of secular non-communication. As in the case of the red deer in the 60's the idea is to boost income, given the relative decline of the cork industry, although here there are some advantages - this animal is a magnificent bio-indicator of the forests and has a minimal impact on vegetation, helping the cork oak to regenerate. And from the point of view of hunting there is its "denomination of origin" as the *Andalusian roe deer*. On the negative side, however, the increase in value of this type of roe deer is encouraging estate owners to import animals from elsewhere, less well adapted to this xerophilous climate and which mate with the local population. This is something similar to what happened with the red deer in the 60's, although it is not so obvious because allochthonous roe deer can only be distinguished with the naked eye in summer when their grey coat changes to a reddish col-

The Roe Deer (Corzo), very common in The Alcornocales forests, is not a subspecies, although it is currently referred to as the *Corzo morisco*, probably for commercial reasons.

Since the early 1990's Deer have
made a great comeback and are
now an important source of income
in the Cork Oak forests.

our. Recently the *Estación de Referencia del Corzo Andaluz* was set up to protect this ecotype, by certifying its origin, breeding and reintroduction on estates which request it.

In the years during which there was an attempt to "improve the quality" of hunting trophies on the Almoraima estate, when it was managed by Rumasa, they also imported fallow deer (*Dama dama*), moufflons or mountain sheep (*Ovis musimon*) and Spanish ibex (*Capra pyrenaica*)! This was to attract hunters by offering easy, spectacular trophies and to make the cork oak forests more profitable. The result could not have been more disastrous since both moufflons and fallow deer prevent the regeneration of the forest because of the enormous pressure they exert on the vegetation and, to aggravate the situation, they have proliferated all over the Park, well beyond the limits of La Almoraima. These herds of undesirable herbivores are very difficult to cut down, unless serious action is taken, and represent a grave danger to the already debilitated cork oaks.

The scrub is also the ideal habitat for the Egyptian mongoose (*Herpestes ichneumon*), the only representative of the mongoose family in Europe. Its body is perfectly adapted to life in the scrub, being elongated with long hairs, a long snout and short ears and legs. No fossils of it have been found in Iberia and, therefore, it may well have been introduced by the Phoenicians or Arabs to catch rats. Thanks to its varied diet of invertebrates, reptiles and rabbits it has not been affected by their respective diseases as much as has been the case of other predators like the lynx, which, as a result, is almost extinct.

Another predator and perhaps the most hated, persecuted and damned is the red fox (*Vulpes*

vulpes). This is "divine punishment" in the form of a wild animal for having exterminated the great predators (wolves and lynx) which kept the fox population in check. Traps, snare and poison seem to have no effect on the fox, there are always more where they came from and it is as if the greater the pressure they suffer, the more cubs they breed and all the more difficult it is to get rid of them. Their omnivorous and opportunist diet enables them to colonize different biotopes and you can even see them foraging in rubbish bins in many towns. According to some hunters they are the *sole* culprits of the fall in game and they are blamed for the death of domestic pets too.

The genet (*Geneta geneta*) is rarely seen because it is an animal of nocturnal habits. Although it is present in scrubland its usual hunting ground is the forest, shinning up trees in search of sleeping birds; it is also the scourge of rats and mice. As with the mongoose, it is African in origin and was introduced by Phoenicians and Arabs as an expert rat catcher.

Finally we have that great musteline, the badger (*Meles meles*), an omnivore and above all a nocturnal vegetarian in these forests, where we also find the weasel, stone marten (*Martes foina*), polecat (*Putorius putorius*) and the "invisible wildcat" (*Felis silvestris*), whose presence has not been confirmed by any forest ranger. Nor has the lynx (*Lynx pardina*) although there are recorded sightings of it at the Pico del Montero near Alcalá de los Gazules in 1991 and at the Cortijo la Vega in Algar in 1993. In recent years there have been no further sightings and perhaps the most reliable proof of its existence in Los Alcornocales was the specimen shot on the Zanona estate in Los Barrios back in 1980.

Two jewels of Wildlife live in the scrub, namely the greater white-toothed shrew (*Cro-

cidura russula) and the pygmy shrew (*Suncus etruscus*). Those of us who have been fortunate enough to observe them can never forget their electric movements, their hyperactivity and aggressiveness. They eat their weight in prey each day and their heartbeat is faster than that of a 100-metre sprinter. The garden dormouse (*Elyomis quercinus*) is less common and more usually found on agricultural land and in pine woods.

The absence of the wild boar (*Sus scrofa*) is surprising as both the vegetation and the lie of the land with its wide plains are ideal biotopes. It may be due to the large number of pigs escaped from farms into the wild, which have bred with wild boar. Indeed it is usual to see striped piglets which later, in adulthood, look like domestic pigs. The uncontrolled spread of these pigs is the cause of the ever-present disease known as African swine fever which is secularly endemic in the area.

Another "missing" species in the area is the wolf (*Canis lupus signatus*) – the last specimen was officially shot in the Montes de Propio in Jerez in 1918, although in the 1980's there was one living on the Almoraima estate. During that time the rangers found half eaten deer and the marks, which they did not recognize, of a "large

dog". The wolf was shot during a hunt, but the affair was hushed up so as not to alarm the general public. Apparently the hunter had the head of the last wolf in Los Alcornocales stuffed.

Among the birds of prey to be found in the scrubland and forest are the Short-toed Eagle (*Circaetus galicus*), Booted Eagle (*Hieratus pennatus*), Buzzard (*Buteo buteo*), Goshawk (*Accipiter gentiles*) and Sparrowhawk (*Accipiter nisus*). We shall deal with them presently in Chapter VI.

The Short-toed Eagle nests in cork oak trees at the edge of the forest, hunting reptiles in the meadows. It arrives in mid-February and quickly occupies its territories of previous years. A common sight is to see a pair chasing away an intrusive young eagle, possibly one which has returned to the area where it was bred in a previous year. A single chick is born in early May and by end July it can fly, although it will stay with its parents until September when they cross the Strait. It is abundant in the Park.

Perhaps the Booted Eagle is the most numerous bird of prey in the cork oak forest and its chicks hatch in early June. It is a skilled hunter, preying on a wide variety of animals like young rabbits, little birds, pigeons etc, and even lizards. We have photographs of this eagle with swifts it has deftly caught in mid flight to feed its chicks.

The Common Buzzard is maybe the least abundant raptor in Los Alcornocales and the first to nest there. In May many chicks are ready to leave their nests when other raptors are still incubating their clutch.

The Goshawk, together with the Sparrowhawk , is the forest raptor *par excellence*. During the year its presence is hardly noticeable until February when they can be seen in nuptial flight. Their chicks hatch in early June and are generally fed on birds from the forest, while those in

The Garden Dormouse is not easily spotted by the casual observer in The Alcornocales Park.

nests near the plains get hares. Only a small percentage of Goshawks take part in migration.

The Sparrowhawk is a partially migrating bird of prey which arrives in March and whose chicks hatch in July, coinciding with the hatching of little birds which are their main source of food.

One of the distinguishing features of the raptors which nest in Los Alcornocales is that some species bring forward their egg-laying, possibly so that the hatching of the chicks may coincide with the arrival of migrant birds, while others delay it to be able to prey on the abundance of chicks of other unsuspecting and less expert species.

Eagle Owls lay eggs in March and their chicks leave the nest only at the end of May. Goshawk and Booted Eagle eggs hatch in the

The Golden Eagle can be seen but doesn't nest in The Alcornocales, possibly due to pressure from Bonelli's Eagle, which does.

first week of June and Peregrine Falcons are still laying in April. The Short-toed Eagle's chick hatches in early May, as it does in other areas, and coincides with the appearance of reptiles as the atmosphere warms up. A detailed study of the phenology of laying, hatching and first flights could be an intriguing subject for a detailed study.

In the early 1920's Irby found several nests of Golden and Spanish Imperial Eagles near to what is now the dried up lagoon of La Janda. There are unconfirmed reports that a pair nested in the *sierra* of Ojén in 1981. When we spoke to a ranger in 1983 we were told that "some people from Sevilla came and took the two eagle chicks away" (sic). Young birds are rarely seen and when they are spotted it is during autumnal migration.

The Golden Eagle (*Aquila chrysaetos*) winters in the area. We have seen them catching Lapwings (*Vanellus vanellus*) on the plains of Tahivilla, near the Laja de Aciscar, but not once during the breeding season.

Other medium-sized resident birds are: the Great Spotted Woodpecker, Wryneck, Golden Oriole, Jay, Wood Pigeon etc. Smaller birds are: the Tree Creeper, Nuthatch (a recent colony), Great Tit, Blue Tit, Robin, Wren etc.

Reptiles are genuine representatives of the African continent and have an ideal biotope in the scrubland. Trapped in the valleys they shun the higher parts in order to avoid the cold temperatures. High mountain chains have an isolating effect and this is quite evident in Morocco where, behind each high mountain range you find different species of reptiles. Here they are a source of food for large raptors and have generally taken the place of the rabbit, which is one of the main reasons we still have an acceptable population of raptors. The viper (*Vipera latasti sp gaditana*) is the only reptile dangerous to man but as it is only active at night it is not usual to come across one in the scrub.

High cliffs and ledges are the home of the supreme predator of Los Alcornocales: the Eagle Owl. In 1987 we monitored a pair with two chicks for some time and found that they preyed on the following species: the Short-toed Eagle, Buzzard, Honey Buzzard, Barn Owl, Scops Owl, White Stork chicks, Redshank, hedgehog,

rat, hare, rabbit, lizard, barbell and five fox cubs from the same litter. It is true that Eagle Owls kill rabbits, but you would have to ask estate owners and hunters the following: How many rabbits would those five foxes have killed, had they not been victims of the grand duke?.

If there is one bird which is synonymous with cliffs it is the Griffon Vulture (*Gyps fulvus*). The Arenisca del Aljibe provides all sorts of hollows and ledges for this vulture to build its nest in. The most populous colony is on the Laja de Aciscar, near La Janda, where 89 pairs were counted in 1994 (Del Junco, O. & Barcell, M. 1994 El buitre leonado en Cádiz). In the Spanish peninsula Cádiz is the second most important province for vultures and the first in free range livestock farming. The number of nests varies from the 89 mentioned down to just two on the numerous cliffs dotted all over the *sierra*. During the last few years there has been a spectacular increase in numbers which we thought may have been affected by mad cow disease, but this was fortunately kept in check and in Los Alcornocales two feeding stations have been set up to guarantee the safety of both animals and humans.

Another carrion bird which nests in cliffs is the Egyptian Vulture (*Neophron percnopterus*), smaller in size and boasting attractive white plumage in adults. It arrives in Los Alcornocales in March and quickly goes about reproduction. It is the most endangered raptor species in the area and since the 1970's, when poison began to be used against pests, numbers have been going down constantly. La Almoraima had 6 pairs then, but there are now none left and the situation is no better in other parts of Andalusia.

The Peregrine Falcon (*Falco peregrinus*) also breeds on the cliffs and its population could

Opposite

The Griffon Vulture is the carcass cleaner of The Alcornocales. Currently under threat due to proliferation of poison bait in the countryside and wind turbine blades and power lines.

be described as normal. Its main competitor is Bonelli's Eagle (*Hieratus fasciatus*) and where this large raptor nests the Peregrine Falcon does not. It is probably because they compete for food that many pairs of peregrines in the Park nest on smaller rocky outcrops, well away from the large sandstone cliffs and surrounded by cork oaks. We remember one case in particular where, in order to see the chicks we had to squat down because the nest entrance was only 1.5 metres above the ground.

In the absence of Imperial and Golden Eagles, Bonelli's Eagle, together with the Eagle Owl, is the king of predators in the area. It maintains acceptable levels of population and, at the time of writing, there are signs that new territories are being occupied. It prefers rocks which are not so high and near meadows, perhaps in order to avoid the exertion required to carry prey up to great heights to feed its chicks. It is sometimes mooted that the absence of pairs of Golden Eagles is due to the fact that Bonelli's Eagle does not allow them to settle new enclaves. We have a photograph of a nest, currently occupied by Bonelli's Eagles, which at the beginning of the last century was that of a pair of Ospreys, as is illustrated in the book by L. Howard Irby, *The Ornithology of the Strait of Gibraltar*.

Paradoxically the most emblematic birds of the cliffs are not the large predators but two species of swift: the White-rumped and the Little Swift, which we described in Chapter I. In the Park it is now rare to see White-rumped Swifts taking over the nests in rocks belonging to Red-rumped Swallows, as was the case in the 1960's and 70's. They have now opted for man-made constructions like abandoned tunnels, old bridges etc. As for the Little Swift, which builds its own, there is a nest in the Natural Park of the

As the countryside is gradually abandoned by man, the Eagle Owl population is growing.

The Kingfisher can often be seen in and around the streams and rivulets of the Park.

Strait and we found another on the Aciscar estate, near a vulture's nest, in 1987.

Living both in cliffs and in woods is the Tawny Owl (*strix alucuo*), a medium-sized nocturnal predator which nests both on rocks and in tree hollows.

Rivers and reservoirs are also an important habitat, although much less well-known than the land. The main point about the rivers is that they are short, given the proximity of the *sierras* to the sea, but very winding in their upper reaches. The most interesting fauna here are the invertebrates, some of which – odonates - are new species for science. But the jewel of these silver streams is the otter (*Lutra lutra*) and fortunately there seems to be an important population of them. We can confirm that rare is the stream where we have not come across excrement of these magnificent mustelines.

Among feathered species there are Kingfishers (*Alcedo atthis*), which together with Bee-eaters and Rollers, are the most multi-coloured of all Iberian fauna. It is not a very common bird as there is a lack of river banks, essential for nest building. We know of a pair of Kingfishers which nest in a bank on a beach in the Strait and another wintering pair in the port of Algeciras.

In the reptile world we have two tortoise species, the stripe-necked terrapin (*Mauremys caspica*) and the less common European pond tortoise (*Emys orbicularis*), the viperine water

Opposite

The Otter, in spite of the deterioration of many rivers, is growing in numbers and can be seen along the coast of the Strait.

snake (*Natrix maura)* and a total of 12 of the 22 amphibians present in the Iberian Peninsula.

In winter you have a chance to see Ospreys on the rivers. At the end of the 80's and in the early 90's there were six individuals wintering on the river Palmones and three on the Guadiaro. Quite a number for such a large bird. In the spring of 2005 a pair built a nest not far from the Palmones river mouth, but it was blown down by the strong *levante* and with it went the hopes of many ornithologists who were longing for them to breed and start a new cycle of breeding pairs in the area. We are hoping that in 2006 this marvellous raptor will return to the rivers of the area.

What is quite odd is the absence of the Black Stork in the Park, perhaps because of the dearth of fish in the upper reaches of the rivers. Perhaps surprising, too, is that there are no crayfish in the upper stretches of the rivers or in rivers like the Hozgarganta, which remains uncontaminated as far down as Jimena de la Frontera, but the absence of calcium prevents the development of their shells.

An endless host of invertebrates teem in the upper reaches of clean, transparent streams and green meadows: enigmatic water boatmen; acrobatic, voracious dragonflies; dytiscids, expert swimmers like tiny tortoises; translucent shrimps and other tiny beings known only to experts who take great pleasure from their ephemeral existence.

We must not leave out the farms; country houses and other human constructions where a surprising variety of fauna coexist with man, to such a degree that many of these animal species disappear if these places are abandoned. Possibly the most popular is the White Stork (*Ciconia ciconia*) which breeds on farms, farm houses and electricity pylons. In the early 80's there was a colony of 24 nests on the Arraez farm near Facinas. When the farm was abandoned by its owners, the colony left, coinciding with the theft of all the eggs in 1983. The amazing thing is that storks made no attempt to nest there in 1984.

The Common Swallow (*Hirundo rustica*) is perhaps, along with the White Stork, the bird which best tolerates and, as we mentioned, requires the presence of humans in order to nest. After the widespread fumigations carried out in the early 60's in Africa, it is estimated that their population fell by 90% and they are no longer seen roosting in their thousands on electricity wires in towns like Algeciras.

The crusade against rats and mice is led by the Barn Owl (*Tyto alba*) which keeps them in check in barns and granaries. In spite of this laudable work its presence is not always appreciated because of the chicks' habit of calling their parents with a strange hissing sound, which many people associate with the great beyond.

In the town of Algeciras, where the new post office building stands, in the 80's there used to be a house and garden, which for some unknown reason had been abandoned by its owners. It stood on a poorly lit corner where not many people passed. One day I read in the local paper that at night strange noises had been heard coming from the house and the talk was that somebody from beyond the grave was playing a macabre melody. Out of curiosity I went there that very night to hear these supernatural noises for myself. I was surprised to find quite a number of smiling people surrounding the house at a safe distance, just in case! Dark had fallen in the month of July and, as the *ghost*, as it was referred to among the happy bystanders, did not appear, some of them, throwing caution to the wind, approached the garden fence. Suddenly from the grave faces and the odd nervous laugh from those nearest the house, it was clear that something unusual was happening. Policemen and fireman -yes, they were present too – hushed the expectant yet fearful crowd, eager for ghostly apparitions. It was then that I could clearly make out the unmistakeable hissing sound of Barn Owl chicks… The incident ended with three chicks in the hands of the authorities, disappointment among the pessimists and ironic smiles on the faces of the rest.

The Natural Park of Los Alcornocales is the launch pad for many bird species which arrive in the area to cross the Strait of Gibraltar. The species, the arrival dates, the local weather conditions and the time of access when approaching the Strait, all determine whether the birds cross or wait. This is why, for migrant birds, both Los Alcornocales and the Natural Park of the Strait are such important sites for gathering into groups, resting and feeding. In the area it is relatively easy to detect roosting places, some of which are repeated in successive years, whereas others are occupied only for days or for a few hours. However, there are a host of little birds which are difficult to observe as they shelter in the forests and scrubland and their passage is imperceptible.

They are neither animals nor vegetables. They are invisible underground but very tasty morsels. Fungi are so important that without them our forests would not exist and one of our long awaited dishes would be missing from our autumn table. For that reason we have decided, as is only right, to include in this book a short homage to them, written with his trademark sensitivity, by my "brother" Manuel Barcell de Arizón, a mycologist and ornithologist from Jerez.

Fungi in Los Alcornocales

Manuel Barcell de Arizón

In Los Alcornocales the last days of September are drawing to a close. During the Indian summer, temperatures have risen again. The hot sun never seems to stop shining. Animals and plants are exhausted from waiting out this long summer drought. The clayey soils are full of cracks and, deep in the cork oak forest, the ground is dusty, leaves crackle under the feet of stags looking for clearings, where they proclaim their supremacy over their harem: it is the rutting season. Besides this, there`s a strong *levante* wind from the east eliminating the last traces of humidity left in the streams of our forests. This hell seems never ending.

Suddenly, one evening when the sun is setting, the wind changes direction. A southwesterly blows clouds in from the Atlantic, covering the sun. At dawn the next day a few timid drops of rain bounce off the dry leaves, gently at first

Paloma Solera with Fungi picked by her parents and friends in the sierras near Jimena de la Frontera.

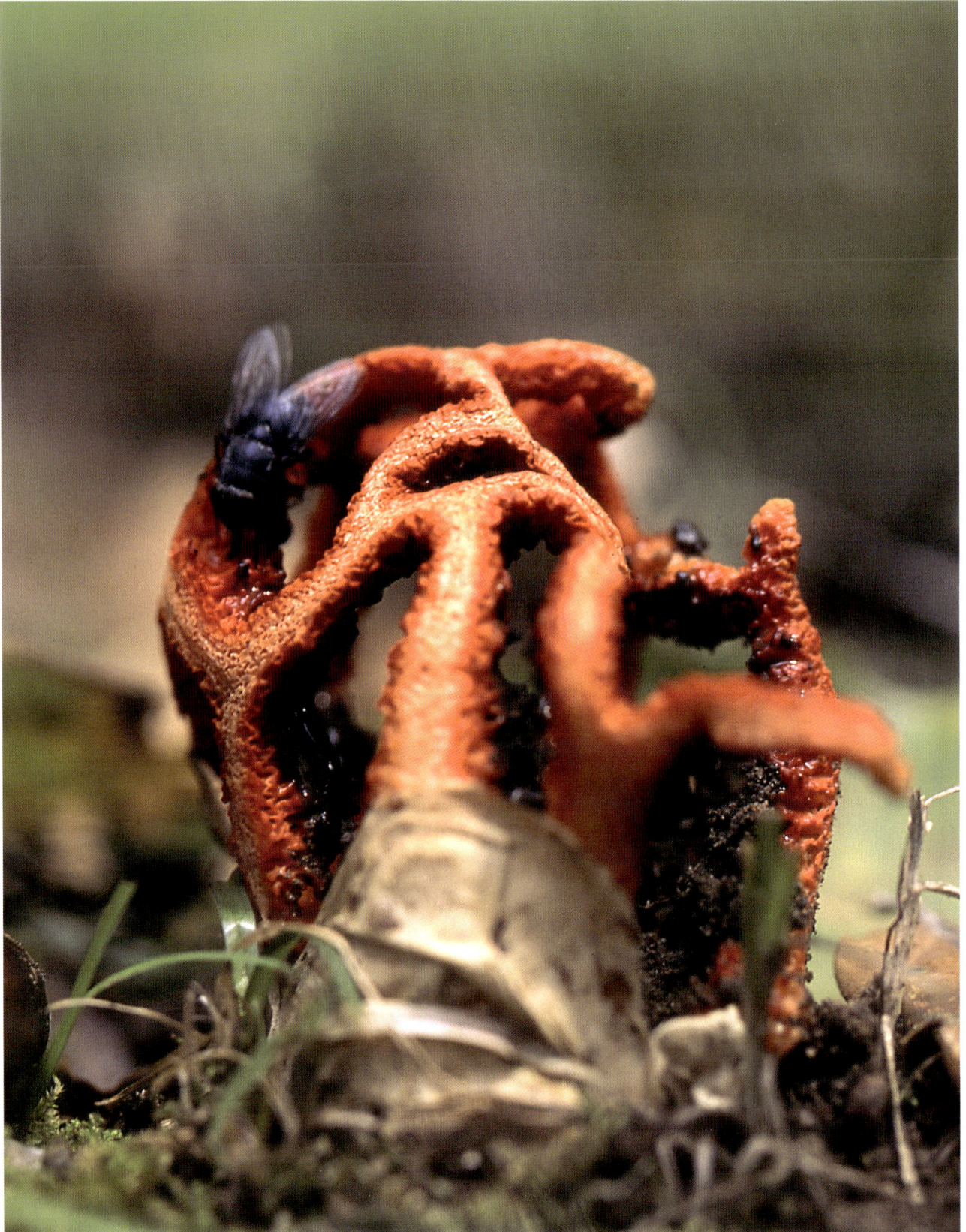

and then more forcefully... the first autumn rains have arrived. Hope is returning to the cork oak forest. A new life cycle is beginning.

Several days have gone by since the first rains. A miracle has occurred in the woods; the ground, the dry tree trunks... everything has changed. A magical world of multicoloured shapes has taken hold of the wood, the first fungi of the season have appeared. In fact, mushrooms are simply the fruit of a fungus whose mycelium, the basic organism, remains underground waiting for the ideal atmospheric conditions in order to emerge.

Depending on their relationship with the forest mushrooms can be separated into three groups. Some are parasites such as *Armillaria mellea*, which exploit the weakness of an old cork oak and feed on it. Others are saprophytes, like the *Pleurotus ostreatus* or oyster mushroom, which lives and feeds on dead trees (a cultivated variety can be bought in the markets in this area). Finally, there are those like the *Amanita caesarea* or egg yolk mushroom, *Boletus aereus* or *Macolepiota procera*, the parasol mushroom, which form mycelia: in other words their mycelium establish a close relationship with the roots of trees and mutually benefit from each other. The tree gets mineral salts and the fungus gets sugars which it cannot produce by itself. This union is obviously very beneficial to our forests.

One of the most popular Fungi is Caesar's mushroom, which, together with the chanterelle, has become a popular dish on local menus.

Opposite

This Stinkhorn mushroom, which really stinks, attracts flies that spread its spores throughout the forest.

The autumn and, to a lesser extent the spring, is the season for mushrooms in Los Alcornocales. The cold winter stops their development and the lack of humidity in the summer makes them disappear.

People in this area have not traditionally gathered and eaten mushrooms. It is only in the last thirty years that mushroom gathering has supplemented the income of some inhabitants of the Park. The chief variety found is the chanterelle (*Cantharellus cibarius*), which people started to gather in Jimena de la Frontera and the custom has now spread to all the villages in the centre and south of the Park. It is a pity, however, that most of the harvest leaves the area to be sold in other parts of Spain and even abroad, so that our area benefits neither economically, nor socially nor gastronomically from one of its own products.

It is getting towards the end of October. For some instinctive reason, our young Griffon vulture feels the urge to abandon the rock surrounded by oak trees where its parents built their nest and where he came into the world. He must set out on the journey to Africa, to his winter quarters. Looking at the multicoloured ground of the forest he takes off and flies away. When he returns in spring, these colours will have disappeared and everything will be green: the life cycle is there in all its splendour. Mushrooms, for now, are a hidden dream.

Southern art cave painting

Although not the subject of this book, but to end this chapter devoted to Los Alcornocales, it is only fair and, we believe, in the general interest, to touch briefly on the subject of rock paintings (pre-palaeolithic and post-palaeolithic), also called southern art in its multiple forms of paintings and engravings in shelters and caves in both the Los Alcornocales and Strait Natural Parks and further afield in the whole Area of the Campo de Gibraltar.

We are referring to prehistoric drawings on walls and ceilings in caves, hollows and shelters, mainly in the sandstone of the Areniscas del Aljibe, gouged out by the erosive action of the wind, and which have lasted until the present day in differing degrees of conservation. They date from the Upper Palaeolithic or Solutrean, through the Neolithic, Mesolithic and Bronze Age periods until the Iron Age.

The majority of the paintings are typical of the early Bronze Age sketch style. One can see arms, helmets, and even sailing vessels (Laja Alta in Jimena). These sketches led to early forms of writing and it can be surmised that certain figures represent concepts. Chronologically, the most recent paintings contain signs that may be Christian in origin.

The engravings in the Cueva del Moro or Moor's Cave, dating from The Upper Palaeolithic, were discoverd by that tenacious and untiring Andalusian investigator, born in Germany but now resident in the area, Lothar Bergman. Because of the alarming deterioration of this sanctuary of Palaeolithic art at its southernmost point in Europe and the inefficiency of the government in preventing vandalism, Lothar went as far as holding a hunger strike in the Moor's Cave. His was certainly a very important discovery, as it puts back the origin of parietal art in the area to a time when the famous figures in the Cueva de Altamira had not yet been painted.

According to various authors, these caves were a kind of sanctuary - Breuil,H. & Burkitt, M.C. 1929 (Rock Paintings Of Southern Andalucía. A Description of a Neolithic and Cop-

per Age Art Group), Hernández Pacheco, E & Cabré, J. 1914 (Avance al Estudio de las Pinturas Prehistóricas del Extremo sur de España), Topper, U & Topper, U. 1988 (Arte Rupestre en la Provincia de Cádiz) – and the relationship between these caves or sanctuaries with paintings and anthropomorphic tombs is very significant. Except in appearance these tombs seem to bear no relation to those in the north east of the Iberian Peninsula, which date back to the 10th and 11th centuries, since experts date them as far back as the late Bronze Age (1000 B.C.) up to Palaeochristian times (700 A.D.) (Sazón 1993).

Fragment of a cave painting in Bacinete (Los Barrios) depicting a fox.

This heritage of nearly 180 caves with paintings, a legacy from our ancestors, ought to be included in UNESCO's list of World Heritage Sites as requested by AGEDPA on its website. This is an association which is fighting to preserve this extremely important heritage and is constantly looking for new sites with paintings.

There is ample room, for any one with energy, to work out a great deal more information on the birds of the Straits.

L. Howard Irby
The Ornithology of the Strait of Gibraltar, 1895

The Rock of Gibraltar

Dr. John Cortés

Introduction

A community which dedicates postage stamps, banknotes and even coins to Nature must be one which delights in showing off its natural wealth to the world. Coins depicting dolphins were found long ago in the Roman ruins of Carteia and perhaps that is why successive generations who have settled in our area and been so impressed with our natural wealth have wanted to celebrate it in this way.

The people of Gibraltar, who have lived on the Rock for three centuries with their own special identity, economy and way of life, have had flowers and birds on their stamps since the sixties, when a Barbary Partridge, a Blue Rock Thrush and the beautiful *Iberis gibraltarica* −such an apt Latin name: *Iberian* and *Gibraltarian* − revealed Nature in the Strait to the four corners of the Earth wherever their mail was sent. Since then there have been orchids, many other types of plants and birds such as the Black Kite or the Peregrine Falcon, even the humble rabbit.

On its coins you can see partridge as well as dolphins, the *Iberis* once more, and its most recently issued banknote depicts Audouin's gulls in flight. And let's not forget the monkeys, the famous Barbary macaques of Gibraltar. It could well be that Gibraltar adorns these long-lasting items with its flora and fauna because, although small, it has a lot to offer and it wants the world to know it.

In the English-speaking world, which in Gibraltar blends with the Andalusian, when the name of the Rock is mentioned, people immediately associate it with monkeys and birds. In Spain, despite its geographical proximity, people are much less aware of what Nature has to offer on this famous rock.

The eastern face of the Rock of Gibraltar and the mountains of the kingdom of Morocco in the background at dusk.

The Rock has been a focus of attraction for ornithologists and naturalists for a very long time. At a time when people still believed that swallows wintered at the bottom of lakes, observations in Gibraltar were crucial in convincing the scientific world about the great exodus and return of birds in their incredible annual cycle. The Reverend John White, an army curate in Gibraltar, was the brother of the famous naturalist Gilbert White, author of "The Natural History of Selbourne" (1788). John White told his brother about his observations and on February 12th 1772 Gilbert wrote to Daines Barrington *"We must not, I think, deny migration in general; because migration certainly does subsist in some places, as my brother in Andalusia has fully informed me. Of the motions of these birds he has ocular demonstration, for many weeks together, both spring and fall; during which periods myriads of the swallow kind traverse the Straits from north to south and from south to north, according to the season. And these vast migrations consist not only of the swallow kind, but of beebirds, hoopoes, Oro pendols or golden thrushes &c., &c., and also many of our soft-billed summer birds of passage; moreover of birds which never leave us, such as the various sorts of hawks and kites."*

White`s fascination with migration and the birds of the area in general was shared by other British naturalists such as the above-mentioned Howard Irby, Verner and Lt. Reid, whose writings were recently discovered.

Doctor Ernest García, a Gibraltarian ornithologist currently living in Great Britain, who was both my inspiration and guide at the beginning of my career as a naturalist, located some manuscripts written by Lt. Philip Saville Grey Reid in 1871. Reid makes interesting descriptions of the ornithological fauna of Gibraltar at that time, such as the pair of Eagle Owls near Middle Hill, the Egyptian Vultures, Bonelli's Eagles and Ospreys. There were also a large number of nests of Alpine Swifts and Lesser Kestrels on the north cliff, numbers of Rock Doves which flew daily to the fields at the north end of the Bay and Black Wheatears on the eastern and southern slopes of the Rock.

The last great character of British ornithological tradition was General Gerald Lathbury, Governor of Gibraltar, who published several pieces of work about the birds of Gibraltar between 1968 and 1970. Sir Gerald was a colleague in the field of a young Gibraltarian named Ernest García, the friend I mentioned above, who organised the first ornithological trips which I went on at the age of 12 and from which the GONHS developed.

Position of the Rock of Gibraltar :

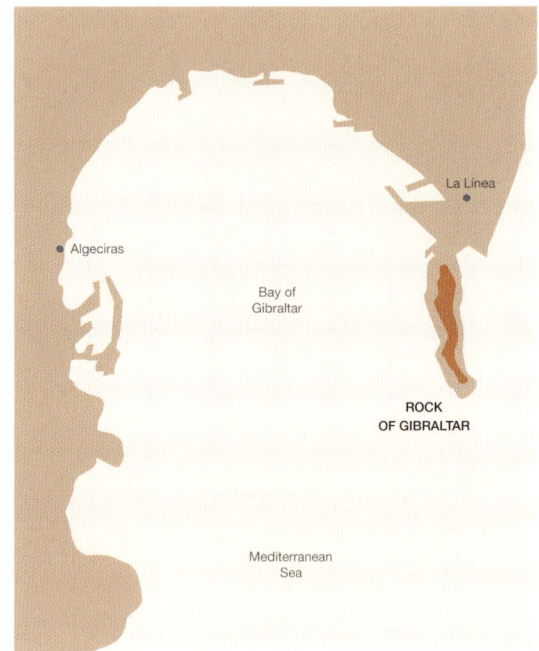

Geology

The Rock of Gibraltar, also known locally as "The Rock", is made of Jurassic limestones which contrast both in chemical composition and morphology to the sandstone which makes up most of the rest of the north shore of the Strait.

This limestone composition is key to the existence on its slopes of a type of vegetation which, although very similar to that in the surrounding area, has some important distinguishing characteristics. One of these is the total absence of heather and the extreme scarcity of rockroses. Another important difference is that, having a lower height than the hills on the other side of the Bay and with the limestone being porous, no water accumulates on the surface. Thus there are no narrow streams or rivers and therefore no rhododendrons or alders.

Vegetation

The wild olive is the dominant plant species and is certainly the most important shrub in the scrub of Gibraltar. Together with *Rhamnus, Osyris, Pistacia, Smilax* and *Clematis* they form very dense scrubland, although their development is relatively recent.

In the XVIII century authors such as Portillo and Ayala described Gibraltar as leafy and green with orchards and trees. Agriculture was common although the olive tree was not cultivated because oil was imported from other areas in exchange for fish, obtained in abundance from local waters and for which the city was famous all over Andalusia.

But everything changed when Gibraltar was taken in 1704 by combined Dutch and English forces. The troops were besieged and, short of fuel, they cut down trees and bushes and even burnt the roots, leaving scant vegetation of poor

quality. This was kept down by goats and cattle so that the Rock looked very similar to the grazing lands to be seen today in much of the countryside in Cadiz.

In more recent times, at the beginning of the XX century domestic herbivores have been excluded from the Upper Rock and there has been a resurgence of the scrubland described above.

Vegetation in pastureland, with the presence of orchids, leguminous plants and a wide variety of flowers, is of much greater diversity to that of the dense scrub, but today this is only found in firebreaks and at roadsides. In the firebreaks you can find interesting specimens such as *Iris filifolia* or *Gladiolus comunis* but this habitat is under threat by two invading species: the native *Acanthus mollis* and *Chasmanthe floribunda* from South Africa. Other foreign invaders are *Senecio angulatus*, a creeper, also South African, which has spread from gardens to the Nature Reserve. On the coast another South African plant, this time a vigorous hybrid of *Carpobrotus edulis* and *C. acinaciformis* is present, although, because it is sterile and does not yield fruit with fertile seeds, it can be easily eliminated by being pulled up and burnt.

The development of the scrub during the last century has affected the fauna on the Rock. Species typical of open countryside or low vegetation such as the ocellated lizard, the Stonechat (*Saxicola torquata*) and the Dartford Warbler (*Sylvia undata*) have disappeared, replaced by scrubland or woodland birds such as the Blackcap (*Sylvia atricapilla*) and the Blue Tit (*Parus caeruleus*). Birdlife on the Upper rock is now a mixture of scrub and woodland birds due to the way the vegetation has developed, the existence of pines planted at the beginning of the XX century and the proximity, especially in the south, of

View from Upper Rock of the Strait of Gibraltar with Jebel Musa mount standing out on the moroccan coastline.

large gardens like the *Alameda* and *The Mount*. So you can find Sardinian Warblers (*Sylvia melanocephala*) together with Blackbirds (*Turdus merula*), Wrens (*Troglodytes troglodytes*) plus a few Great Tits (*Parus major*). Many of the Aleppo pines (*Pinus halepensis*) and stone pines (*Pinus pinea*) succumbed during the drought of the early 1990s (see an analysis in the excellent work in *Almoraima* by Keith Bensusan and Charlie Pérez). These trees usually attract birds such as Bonelli`s Warbler (*Phylloscopus bonelli*), Iberian Chiffchaff (*Phylloscopus ibericus*), Short-toed Treecreeper (*Certhia brachydactyla*), occasional Coal Tit (*Parus ater*) and Crossbill (*Loxia curvirostra*), either on passage or for stays of several weeks. The Tawny Owl (*Strix aluco*) is becoming more common and may indeed be colonising the Rock's copses and gardens.

The Collared Dove (*Streptopelia decaocto*) has been resident in gardens since 1991 and, together with the Blackbird, has become one of the most common birds of this habitat, replacing the humble Sparrow (*Passer domesticus*) which, as in so many other places, is disappearing. You can hardly ever see one in the city where however there is an abundance of domestic pigeons feeding on scraps of food thrown to them.

But Gibraltar is much more than just scrubland. Perhaps the most important biotope is the sand dune on the east side. The sand has accumulated there thanks to the easterly *levante* winds which have carried it from the ancient plains in the east and formed a gigantic dune about 300 metres high. It is here, as we shall see later, that the original vegetation is being re-established. With species such as *Malcolmia littorea*, *Medicago marina*, *Delphinium nanum* and *Pancratium maritimum* and, endemic to the area, *Ononis natrix natrix ramosissima*, the influence

A yacht entering the Strait on a spring evening in 2005. On the right the lighthouse at Europa Point, on the left the atlas mountains.

of the sea is obvious although among this coastal vegetation there are also examples of *Ferula tingitana*, *Ephedra fragilis*, and even the wild olive (*Olea europea*).

Ferula tingitana and *Ephedra fragilis* are species not often seen in the area surrounding Gibraltar. *Ephedra*, to which is associated a curious type of coleopteran, *Yamina sanguinea*, according to my colleagues Charlie Pérez, Keith Bensusan and Juan de Ferrer, is also present in the sands next to the river Palmones estuary, but apart from there this shrub has not been recorded anywhere else in the surrounding area.

Other rare species in the area are *Succowia balearica* and those beautiful leguminous bushes with the fragrant yellow flowers *Coronilla valen-*

In the foreground flowering *Iberis gibraltarica* on Upper Rock. One of the most beautiful flowers in the area.

tina. The latter is found in less dense scrubland, flowering in January and February just before the Rock's other two similar species, *Calicotome villosa* and *Genista linifolia*.

Any visitor to Gibraltar who has seen it from afar and expects to find just another large hill will be surprised at the number of nooks and crannies, crevices and precipices in this ancient Rock. And in these hidden corners untouched by man and his animals, some unexpected plant species still survive. The famous bay tree (*Laurus nobilis*), a proud resident of the nearby Natural Park of Los Alcornocales, is also a native of the cliffs of Gibraltar. Mediterranean fan palms (*Chaemerops humilis*) protected by their sheer inaccessibility, are of considerable size with trunks up to 2 metres, giving a tropical touch to the blue Mediterranean background.

A botanical species which grows very well here is the Gibraltar sea lavender (*Limonium emarginatum*), an endemic littoral plant, which, although present in other shores on the Strait, usually grows larger and leafier on the Gibraltar limestone.

Indeed it is the plants growing on cliffs and other rocky areas which are peculiar to Gibraltar. Some of them, such as *Thymus wildenowii and Iberis gibraltarica*, might still be found in the mountains of the Tangier Peninsula in northern Morocco, where they have been gathered in the past. *Thymus* was located some years ago near San Roque but has since disappeared. This Gibraltarian thyme has whitish flowers, is not very aromatic and grows over rocks. It is not as colourful as *Iberis* which blooms splendidly from March to May on cliffs and in the more open scrub. Its flowers, which go from white to pink and almost lilac, adorn the Rock in spring.

It is much more difficult to find *Saxifraga globulifera*, a little plant with downy, dark green to red leaves and small, delicate, white flowers. There are similar plants on nearby mountains, as in the Sierra Blanca in Málaga, but some botanists consider them to be an endemic subspecies.

The most emblematic flower is, without a doubt, *Silene tomentosa*. Classed as endemic to the Rock, it was photographed in 1987 by my friends Leslie Linares and Arturo Harper, botanists and photographers who co-wrote with me

In Spring many invertebrates both polinate and feed on colourful flowers.

Silene tomentosa, an endemism rediscovered in the 1990's and now being replanted to avoid its extinction.

The Flowers of Gibraltar. But then it disappeared although we searched for it in vain every year in April and May.

On the 4th May 1994 when we had just given up after another year's fruitless search, Leslie, Arturo and I were walking back down the hill when we suddenly saw a plant covered in pale pink flowers… And I said: "That's it, isn't it?" "Yes, it is" replied Leslie quite calmly….. until we realized what we had just said. We located three plants that day. A few days later I sent some cuttings to the Royal Botanical Gardens at Kew in London and several weeks later I went to collect the seeds.

We had found the plants just in time. In July the long drought, which lasted from 1993 to 1995, finished off the three plants, and the ones which exist now, including those which have been used for a modest replanting scheme, those donated by the Gibraltar Botanic Garden to the one in San Fernando and those in the Eden Project in Cornwall in England, all descend from them. Luckily these three plants retained great variability and among their descendents there are plants with flowers ranging from pure white to strong pink. They all have an incredible aroma – it is sweet, reminds one of the sea and gives me great satisfaction.

The cliffs on the Rock are quite spectacular. They stretch from the north side, with a sheer

Opposite

Snapdragon, which favour limestone soils, can often be seen growing on the roofs in the city.

drop of 300 metres, to the isthmus which joins the Rock to La Línea. Then they go in a north-south direction to form a high barrier which rises to 426 metres at to form the westernmost limit of the Mediterranean Sea.

Fauna

In prehistoric times Iberian ibex and colonies of Griffon vultures lived on these cliffs – as can be seen from the fossils found in the many caves existing on the Rock.

These caves have been, and are still being studied for many years by palaeontologists from all over the world. In one of them, Devil's Tower at the foot of the north cliff, the skull of a Neanderthal woman was discovered in the mid XIX century a few years before a similar one was discovered in Germany. As it was not immediately recognized as a new form of human being it was stored away and only came into prominence was after the species had been named *Homo neanderthalis*. With closer observation or greater scientific knowledge, perhaps it would have earned the name of *Homo gibraltariensis*. When Neanderthal man lived in this area during the glacial period, the level of the sea was much lower and a sandy plain stretched several kilometres to the east of the Rock, possibly with vegetation similar to that of present day Coto de Doñana. Among the remains found in the caves, corresponding to that period, are those of deer and ibex, small mammals and many bird species still seen today in the area such as the Shag, Rock Dove, Griffon Vulture, Eagle Owl and Swallow.

Both the Neanderthals and their successors, modern Man, brought food into the caves. That is why there are so many remains of animals and shells and even tuna fish, showing that they used the sea as a resource. There are also tools made of stone and bone.

The fauna of the area has seen many changes and there are remains of species no longer present in the area such as the White-tailed Eagle and even more interesting ones like the Francolin, a species found in Spain until the mid XIX century. There have been more recent discoveries of ornithological interest, like the remains of the Iberian Azure-winged Magpie (*Cyanopica cyanus*), refuting the theory which says that they were introduced from the Orient centuries ago, and the Northern Bald Ibis, thereby giving historical backing to the attempts to reintroduce this species in the area.

The varied population of nesting birds on the cliffs has survived until very recently. At the beginning of the XX century Egyptian Vultures, Bonelli's Eagles and Ospreys still nested there but gradually disappeared perhaps as a result of military activity during WWII or blasting on the Rock to obtain material for the construction of the airport runway.

However, the community of birdlife nesting on the cliffs has not been completely lost. The melodic song of the Blue Rock Thrush can still be heard in spring. But it is the Peregrine Falcon which has become the emblematic species of the cliffs of Gibraltar. Up to six pairs nest every year in Gibraltar and they specialize in hunting, among other birds, the thousands of migrating passerines which venture into their territories. Every year between 13 and 20 young falcons make their first flights in the skies around the Rock of Gibraltar.

Another small daytime raptor also survived the conflicts of the XX century. The English ornithologist, Colonel Howard Irby, who wrote the *Ornithology of the Strait of Gibraltar* in 1875,

On the sheer cliffs of the Rock six
pairs of nesting Peregrine Falcons
breed early to feed their young in
spring with migrating birds.

The Lesser Kestrel population has diminished
considerably since Irby reported several
hundred on the east face of the Rock in 1875
and is now down to about eight pairs.

referred to a population of several hundred Lesser Kestrels on the north face of the Rock. This attractive species used to nest in orifices in the cliff and would hunt in the marshlands and sands of the isthmus and the countryside around La Línea. In recent times as has happened in other parts of the region, the population has diminished considerably and there are now just eight or nine pairs left, which have to fly further and further away for food and in so doing probably use up more energy than they recover by catching some large cicada or small lizard.

On the other hand, the Common Kestrel, which only nested occasionally in Gibraltar and was considered more of a migrant than a resident, has colonized the city's buildings since 1987 and often feeds on, among others, canaries from cages on balconies!. There are nine pairs nesting in Gibraltar, most of them still on cliffs.

Since the latter part of the XIX century the presence of the Eagle Owl, another former resident has been occasional or just a mere suspicion. This suspicion, however, was confirmed in 2004 with the presence of at least one calling male – and the release of a female from the GONHS (Gibraltar Ornithological and Natural History Society) rehabilitation centre. The presence of this great nocturnal predator may explain the noise often heard in the early hours coming from the large colony of seagulls – or the lack of falcon chicks in recent years in the nests nearest to the crevice where the owl was located.

The return of a locally extinct species always gladdens the heart of a naturalist; especially when this reappearance is by totally natural means. This was the case of the Raven, another former member of Gibraltar's birdlife com-

A pair of Ravens is now trying to breed on the Rock, but some physical defect in the male prevents him from mating, so as yet they have no descendents.

munity. Ravens nested in Gibraltar until 1972. From then on only one member of a pair was seen until it disappeared in 1974. Since then Ravens have only been occasional visitors and the decrease in numbers has been noticeable in the surrounding area. But from 1999 onwards a new pair of Ravens appeared on the Rock. They have built nests since 2001 but so far without producing offspring.

The characteristics of the cliffs change the nearer they are to the sea. In the south-eastern part of the Rock the vertical limestone face goes right down into the Mediterranean Sea. It is here where the only colony of Shags in the area takes refuge from the strong *levante* winds and storms. They too have seen better times; you could often see up to twenty Shags at their favourite resting place in the 60s. But over-fishing, the proximity of the tuna fishing ground at La Atunara,

The Barbary Partridge survives despite coming under pressure from wild cats that catch them.

where many Shags have perished, and pollution through oil spills, have had a negative effect on these fishing birds. Numbers went down to three or four pairs although in the last few years they seem to be increasing slowly.

Not far from the shag nests and in the cliffs and huge sea caverns sculpted by centuries of *levante* windstorms, nest both Pallid and Alpine Swifts. Together with another small population on the north face, they are the only known nesting sites for the Alpine Swift in the whole area of the Strait.

Gibraltar birdlife enriches the surrounding area by contributing species which are not seen beyond its limits. Perhaps the most well-known is the Barbary Partridge (*Alectoris barbara*). This North African species was probably introduced by the British military in the early XVIII century as a game bird for officers of the garrison. It can be seen wherever there is low vegetation, from the firebreaks on the Upper Rock right down to the Mediterranean on the east side.

In the early XX century the population of Gibraltar was increasing and water was in short supply. It was then that a start was made on what was probably – and still is – the most impressive civil engineering project in the area of the Strait. Gigantic artificial caverns were blasted out of the Rock, and the east sides were covered with stainless steel plating, forming an enormous rainwater catchment and storage system. While this guaranteed a supply of drinking water for Gibraltarians, it had a very negative impact on the local flora and fauna. About 160 hectares of sandy slopes with a great diversity of flowers and an animal community which included skinks, scorpions and spiny-footed lizards (*Acanthodactylus erythrurus*) disappeared under a sea of metal. Perhaps the most signifi-

Present-day view of the eastern face of the Rock after removal of the steel sheets.

The new habitat created on the east slope has already been visited by a well-known resident of the Rock, the Barbary macaque or Rock ape (*Macaca sylvanus*).

Mammals

cant loss was the extinction in Gibraltar of the Black Wheatear...

The arrival of new technology, desalination, made these metal plates obsolete. For safety reasons they were removed and, thanks to the GONHS with the support of the Gibraltar Botanic Gardens, work was begun on the restoration of the native vegetation using seeds collected from the sands and dunes of the area. The vegetation is recovering successfully and the reintroduction of the beautiful Black Wheatear cannot be far off.

The Barbary macaque is the only macaque which exists in a natural state outside the Asian Continent and islands. It is endemic in Morocco and Algeria where the population of several thousand is under threat. They existed in Europe in the interglacial periods but did not survive the last glaciation. The monkeys probably arrived in Gibraltar when the British came as there are references in official documents to the importing of animals from Morocco for hunting. The current population is about two hundred split up into six groups of thirty or so animals with a complex social structure. They are fed daily

Opposite

An old photo of the east slope of the Rock showing the steel sheets that covered the dune in order to collect drinking water.

by GONHS staff: fruit, vegetables, wheat and sunflower seeds, which the animals themselves supplement with flowers, fresh leaves, fruit from wild olive trees and other bushes as well as the odd invertebrate for protein.

The groups are divided into families around the matriarchs and have a strict hierarchy in which the dominant animals always have preference at meal times. This hierarchy is hereditary and the young of dominant females are higher up the scale than older animals whose mothers are lower down in the hierarchy.

This species has a very curious way of warding off aggression. It entails carrying a young one in its arms, and the males often do this when they want to approach a higher ranking animal. While carrying a young one of the group it will never be attacked by another member. This almost human characteristic of respect and protection of the young is unique to this species of macaque.

Sometimes, when a group increases to over fifty members, it will split up, always in families and then become nomadic while searching for a new territory. This may take them to the city of Gibraltar where they become a nuisance as they enter homes and food shops looking for something to eat.

This is quite easy for them because, sadly, they are used to the sweets given to them by tourists in exchange for photos. They associate people with sweets which not only harm the monkeys` health – their teeth rot and they get fat – but also make them more aggressive when they don't get what they expect from humans. Besides, the contact between monkeys and humans can easily lead to the transmission of diseases in both directions. It is better to let them eat the food provided officially and what they get from nature, respecting them as just another member of the fauna of the Strait and not treating them as circus animals.

In spite of the presence of these curious primates, the Rock's mammal community is rather poor.

Rabbits (*Oryctolagus cuniculus*), bats and shrews (*Crocidura russula russula*) are the only native species which survive.

While preparing the Management Plant for the Upper Rock Nature Reserve, my colleagues Charlie Perez and Keith Bensusan came upon an unpublished piece of work by the Gibraltarian scholar and historian George Palao which describes a total of 107 caves in a census carried out between 1966 and 1968. Bats were very common at the time. Palao identified roosting places of more than a thousand greater mouse-eared bats (*Myotis myotis*), more than 6,800 Schreiber's bats (*miniopterus schreibersi*), quite a number of pipistrelle bats (*Pipistrellus mediterraneus*) and free-tailed bats (*Tadaridis teniotis*). The large colonies of *Myotis* and *Miniopterus* have now disappeared. In 2004 only one roost of about 200 hibernating *Miniopterus* was present, although now and again some *Myotis* could be seen, especially in the Botanic Garden. My colleague Tony Santana, a great enthusiast of these mammals, has inspected the rings of some of the *Schreiber's bats* and found that they had been ringed near Benalmádena in the province of Málaga. There could be several reasons for the disappearance of so many bats. They are known to have been disturbed in their caves, even though they are a protected species, and some species have contracted diseases in other parts of the Iberian Peninsula. But the most likely reason is that as the scrubland has grown these mammals have lost their favourite habitat for catching fly-

Opposite

The Barbary Macaque was introduced by the british, as was the Barbary Partridge, for hunting purposes.

The Shrew is one of those living gems that even expert naturalists fail to spot.

ing insects – clearings and flat pieces of land – and have therefore just left the area.

The other mammals present are, of course, mice (*Mus domesticus*) and rats (*Rattus rattus alexandrinus*). And there is one serious problem: feral cats. There is a considerable and growing number of these, due to their being fed by misguided animal lovers. The worrying thing is that they prey on small birds, even Barbary Partridge and their chicks and you can often find the remains of tired migrating birds like Hoopoes and Turtle Doves, victims of these predators.

There are historical data of two native carnivores, the genet and the fox. The latter disappeared about 25 years ago and attempts are being made to reintroduce it. Foxes are known to control the cat population and will have less impact on other species than the colonies of 20 to 30 cats seen in some parts of Gibraltar. They would also be a welcome help in controlling the seagull population. The possibility of reintroducing the genet is also under study.

Because of the growth of its human population Gibraltar lost all its great herbivores long ago. The last wild boar was hunted in the early XVIII century, and the ibex probably centuries before that.

So plans are being developed to reintroduce the ibex (*Capra pyrenaica*) and to import the Andalusian roe deer (*Capreolus capreolus*). It is hoped that these two species will be able to help in keeping the growth of the vegetation in check.

The few reptiles living on the Rock are of the species commonly found in scrub throughout the region. There are not too many reptiles living on the Rock,being of the species commonly found in scrub throughout the region. There are no adders (*Vipera latasti*) but the horseshoe whip snake (*Coluber hippocrepis*) is quite common. Others are the ladder snake (*Elaphe scalaris*), the southern smooth snake (*Coronella girodica*), the Montpellier snake (*Malpolon monspessulanus*), the false smooth snake (*Macroprotodon cucullatus*), the viperine snake (*Natrix maura*), and the European grass snake (*Natrix natrix*).

As for lizards, the Iberian wall lizard (*Podarcis hispanica*) is certainly the most common and is the species to which I devoted three years of my life, its ecology being the subject of my doctoral thesis. Moorish geckos (*Tarentola mauritanica*) are common but the Turkish or disc-fingered gecko (*Hemidactylus turcicus*) is rarely seen. Anther common inhabitant of the scrubland is the large psammodromus (*Psammodromus algirus*) but the Spanish psammodrumus (*Psammodromus hispanicus*) is inexplicably absent. The ocellated lizard (*Lacerta lepida*) is very rare and absent from most of the territory but both Bedriaga's skink (*Chalcides bedriagai*) and the western three-toed skink (*Chalcides striatus*) are found, mainly in sandy areas. The Iberian worm lizard or amphisbaenian (*Blanus cinereus*) is common too, especially in gardens.

As there are no rivers or streams or any other natural fresh water sources, the European pond tortoise (*Emys orbicularis*) and the Spanish terrapin (*Mauremys leprosa*) are only found in the ponds in the Alameda Botanic garden, where there are also a number of introduced red-eared sliders (*Trachemys scripta elegans*) and small introduced populations of the stripeless tree frog (*Hyla meridonalis*), common toad (*Bufo bufo*) and Iberian water frog (*Rana perezi*).

Migration

We have already seen how owls and ravens have returned after many years' absence. But there are many other birds returning year after year. They are the main subject of this book and of this chapter: migratory birds.

But first of all let us not forget that the Strait is a focus of migration for other animals too. Thanks to the pioneering works of Eric Shaw it is well known that dolphins live in our waters, notably the common dolphin (*Delphinus delphis*), striped dolphin (*Stenella coeruleoalba*) and bottlenose dolphin (*Tursiops truncates*). The passage of several species of whale sees the presence in the Bay in summer, for example, of the minke whale (*Balaenoptera acutorostrata*). Obviously all whales moving between the Atlantic and Mediterranean must pass though the Strait.

When the winds blow from the west Honey Buzzards fly over the city of Gibraltar on both migrations.

At the end of summer and in early autumn Dragonflies invade the shores of the Campo de Gibraltar and are food for Eleanora's Falcons which stop to rest in Gibraltar during autumn migration.

The migrations of tuna fish (*Thunnus thynnus*) and frigate mackerel (*Auxis thazard*) have been known since prehistoric times and these fish have been a source of food for the peoples of the Strait for thousands of years. But overfishing has led to the disappearance of the tuna fisheries of La Línea and Sabinillas, a sad reflection of Man's lack of respect for nature, the results of which, as always, cost us dear.

Less well-known are the migrations of insects. In autumn large numbers of dragonflies arrive in Gibraltar whereas during the rest of the year they are not common. But in some years the most spectacular are the spring arrivals, in March or April, of thousands upon thousands of butterflies. Among the most common is the painted lady (*Cynthia cardui*). These fragile invertebrates can be seen fluttering erratically over the sea towards Europa Point and once on land they cover all available flowers to drink the nectar which gives them fuel to continue their journey. I remember one spring morning on top of the Rock when the arrival of thousands of large whites (*Pieris brassicae*), was reminiscent of a fall of large, delicate snowflakes.

The arrival of the beautiful monarch butterfly on the shores of Gibraltar was not the usual migration. This North American species, resident in the Canaries since the end of the XIX century, joined the fauna of Gibraltar in 1998, possibly when some nursery plants were brought to the area. Right away some *Asclepias curassavica* were planted in the Botanic Garden to feed the larvae of this attractive orange and black butterfly. Since then they can be seen flying around the attractive gardens of the Alameda, especially in summer when other large butterflies like the two-tailed pasha (*Charaxes jasius*), whose larvae feed on the *Osyris quadripartita* in Gibraltar as there are no strawberry trees, and also the swallowtail butterfly (*Papillo machaon*) and the rare and aptly named scarce swallowtail (*Iphiclides podularius*).

But, returning to bird migration, let me have a nostalgic look back to a hot, August afternoon at the very top of the Rock. To the north the dry mountains of the eastern part of Cadiz province from Sierra Carbonera to Ronda; to the west the blue sea of the Bay, with that dark tone it gets from the *poniente* wind, and on the other side of it more mountains browned by the burning sun; then, to my right and 400 metres below, the beaches of Mediterranean Gibraltar with clear deep green waters and hundreds of families enjoying themselves on the shore.

I had climbed up alone that day. I usually came up the steps along the Charles V wall with friends on a cool late summer morning when a hint of September was already in the area and the sun was just appearing above the horizon. But that day I was alone. Throughout the morning, from first light, thousands of Honey Buzzards had streamed in lines over the Rock. At first they flew very, very low, leaving their roosts in the pine and cork oak forests and arriving in Gibraltar before gaining height to cross the Strait. They came from the north, little black dots at first against a background of Andalusian hills, and as they came closer you could see the typical variation in plumage of the species – some almost black, others almost white. And they came from the east too, black dots over the sea, approaching the cliff below me and rising slowly, turning in large flocks while the morning sun began to heat up the sides of the Rock. As the sun rose and it got hotter the buzzards had already gained height inland, so they arrived at the Rock ever higher until the long lines

of birds almost disappeared and could only be seen through binoculars like little ants crossing an endless blue field, with the occasional flash when they reflected the sun's rays.

But it had been a hard day. The wind had dropped and therefore the number of birds too. After a drink and a sandwich and tortilla prepared the night before by my patient mum, I leant back looking skywards, with just a low wall between myself and the precipice on my right.

I fell asleep in the sun for a quarter of an hour or so. But not wanting to miss a single raptor I jolted myself awake and looked up. The sky was empty. Sitting behind the wall I couldn't see what was going on below me to the right. I sat down slowly and my eyes came level with the top of the low wall.

Having left a wood in Cordoba at dawn and flown non-stop, a young Black Stork was slowly gaining height, on the thermals which were rising from the eastern side of the Rock.

This Black Stork and I crossed the summit of the Rock at the same time and for an instant, looked each other in the eye, face to face, eye to eye, just 50 centimetres apart. The startled stork, pulled back abruptly, glided down carried on, on its journey to Africa. I kept my eyes glued to it, amazed at having had such an emblematic migratory creature so near to me.

The most spectacular sight in bird migration, when observed from Gibraltar, is the post-nuptial passage of soaring birds, especially the Honey Buzzards.

The geography of the Rock, a mountain which lies north-south, a unique promontory of high land in the eastern part of the great Bay, is a focal point of attraction for all soaring birds which, when the wind is from the west, drift to this end of the Strait. Thus it is that, depending

Opposite

Black Storks don't often fly over the Rock, but in 1997 some 328 were recorded, an all-time record.

Part of a flock of Griffon Vultures approaching the
Rock, on prenuptial migration, with a prevailing
poniente wind.

on the time of year, thousands of Black Kites, Honey Buzzards and hundreds or dozens of other species such as Harriers, Egyptian Vultures, Ospreys and Booted Eagles mass together over the Rock. This phenomenon caused some young Gibraltarians to become interested in birds as they started to admire them and write down field notes back in the 60s. Four decades of observing migration have given us the opportunity to observe differences in its phenology, something which is now being actively studied. Perhaps the most noticeable change has been the almost complete disappearance of the Common Buzzard as a migrant. In the springs of the decade between 1965 and 1975 sometimes more than 800 were counted. Now this is down to barely a dozen in any one spring. It is not easy to pinpoint the reasons for these changes, although logically one thinks it may be the result of the general increase in temperatures.

The summit of the Rock is one of the few points in the Strait where during the southward migration there is an unobstructed view to the north. It not only attracts the birds – they can also be seen better from here, often from above. Gibraltar also has its special attractions in autumn and the Eleanora's Falcon is one of them, no doubt attracted by the sea cliffs reminiscent of their normal breeding habitat. Most years, one interrupts its journey to spend a few days resting and catching dragonflies and small birds on the Gibraltar cliffs, until it is chased away by the resident Peregrine Falcons.

When the levante blows from the east, conditions are unfavourable for migration over Gibraltar, with one exception. In September Booted and Short-toed Eagles fly over our area. When the winds are strong and visibility is poor these raptors do not cross the Strait but accumu-late in the area. On such days several hundred of these species, plus Egyptian Vultures and even a few storks, coast along to Gibraltar. The Booted Eagles usually stay for a few days and can be seen hunting on the Rock and in gardens, on these damp and misty days.

As the arrival of the raptors in Gibraltar depends on the wind, the total number of each species observed from the Rock depends on the combination of wind direction and the key migration time. Thus, for example, if a levante is blowing at the end of August and early September, in that year you will see very few Honey Buzzards, so that the black kite might be the most common bird of the season. It will always be one or the other. Between 1967 and 2001 the postnuptial migration count was from just over 4,000 up to nearly 13,000 Black Kites and between 7,000 and almost 46,000 Honey Buzzards.

Kites often arrive in Gibraltar in the mornings, flying at medium to low heights and in a

Soaring birds flying northwards over the Rock :

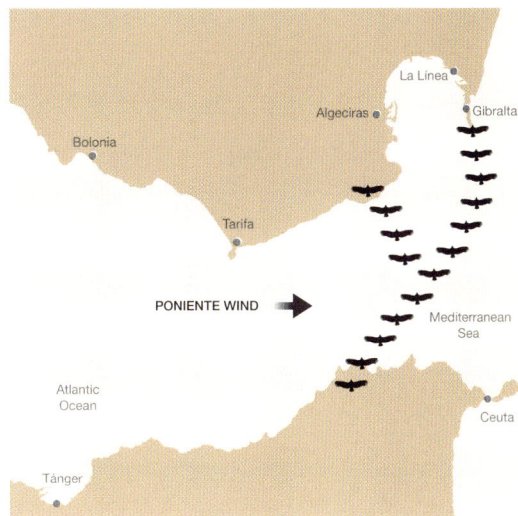

relatively narrow stream. Honey Buzzards, on the other hand, although flying in low early in the morning, by mid afternoon can be mere dots even when seen through binoculars. Honey buzzards move in a broad front, and not just over the Strait area. They can often be seen to the east of the Rock flying north-south over the Mediterranean. I remember sailing in a small boat one September afternoon some 12 kms east of the Rock. Looking skywards, I saw a small flash of light in the sky and, on closer examination through binoculars, I could see a large group of Honey Buzzards flying dead south without the slightest intention of dropping in on Gibraltar or the Strait.

Of the other migratory species which gather in Gibraltar every year only Booted Eagles number more than a thousand, although on occasions Griffon Vultures have also had this distinction.

There are some species which are relatively rare in this sector of the Strait. White Storks are rarely seen in postnuptial migration, in contrast to the large numbers further west. Only when they return northwards after spending time in Africa, can one sometimes see groups of several hundred flying over Gibraltar. This prenuptial migration can start as early as late October and last until April.

The journey south begins in July. During this month black kites and thousands and thousands of Swifts (*Apus apus*) cross the Strait from Gibraltar. In August, as we have seen, there are even more. As well as the soaring birds others flying south at this time are Bee-eaters (*Merops apiaster*), Alpine Swifts and Fan-tailed Warblers (*Cisticola juncidis*), not usually regarded as regular migrants.

Migration continues in September, through October – when the vultures can start appear-ing– to early November. Nearly every November and December a group or two of White Storks turns up, but apart from these, the migration of soaring birds stops for Christmas.

In Gibraltar, especially where soaring birds are concerned, it is quite easy to predict the type of migration according to the date and the type of wind – not only because the birds follow a pattern but also because of years of experience of the observers there. For example, during postnuptial migration, in August, which is Black Kite season, the westerly wind carries them along a fairly fixed route, north-south or south-west, some groups flying towards Punta del Carnero and others directly towards Morocco. When the Honey Buzzards are migrating a strong westerly will make them drift eastwards, as we've seen, well beyond the Rock. If night is drawing near they may change direction and fly into the wind towards Europa Point from where they'll head north, either to roost in trees on the Upper Rock

Soaring birds flying southwards over the Rock :

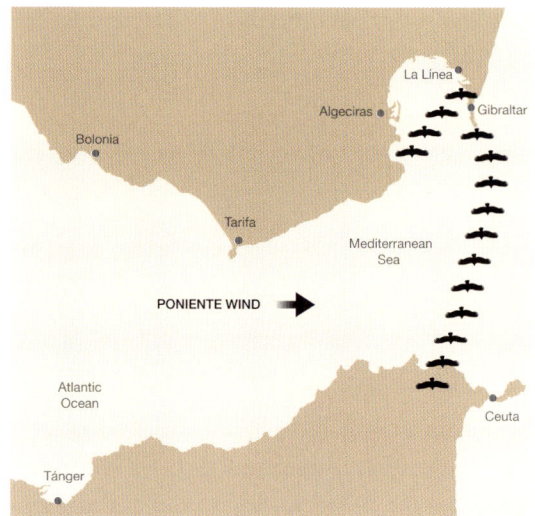

or to continue to the northeast and into some pine or cork oak forest in the area.

The Honey Buzzards may arrive in several groups at once: some out over the Mediterranean, others over the Rock like the kites already described, yet others at low altitude over the city and if the wind drops, a stream will skirt the north of the bay towards the east/southeast. It is a really complicated task to keep abreast of all these simultaneous streams and is something which Charlie Pérez, Paul Rocca, Paul Acolina, Keith Bensusan and other observers, who have been studying raptors in Gibraltar for decades, have managed to become experts at doing.

February is the month when you know that spring has really come, with clear skies, plenty of sun and, of course, westerly *poniente* winds. It's the time when you can expect to see the first kite of the year. February turns to March and March into April and there is a constant increase in the passage of birds of prey. Again it's the poniente winds which bring them towards the Rock. As they fly northwards the Strait has the effect of ironing out the differences between the different species, whose arrival on land does not so much reflect their prenuptial African route but rather the prevailing sea breezes and winds. So in general we can say that when the levante wind prevails birds do not arrive in Gibraltar and when the wind is westerly (poniente) they aim directly for the Rock. Jews' Gate ornithological observatory overlooks the eastern sector of the Strait. Looking out for raptors now is very different from the autumn. With the Moroccan hills in the background and much greener countryside than during the autumn passage, little black specks can be distinguished in the distance, each with its own way of moving and with a distinctive shape and "jizz" which allows them to be identified from a long

way off. Although the number of passage birds is lower than in autumn, several thousand kites and, from late April to mid May, Honey Buzzards along with hundreds of birds of other species can be counted from the Rock of Gibraltar.

Once again the most frequent migrants are the Honey Buzzards and Black Kites. Data from the GONHS show a minimum of 2,381 and maximum of 11,841 Black Kites and from 3,859 to 12,667 Honey Buzzards, the numbers again depending on the prevailing winds.

At this time of year the passage of flocks of Swallows, House Martins and finches provides an amazing feat of nature. It was just such a spectacle in spring which inspired the Reverend John White to write to his brother.

However, in spite of the magnificent show put on by the migrating birds, most people in our area are blissfully unaware of it. But they do notice some changes. Many people hear the whistling of the Swifts when they arrive at their nests in spring. Whether they be Pallid in February or Common Swifts in April, their presence is a typical sign of summer. In Gibraltar people call them "Aviones", which is Spanish for House Martin (as well as aircraft!): "The *aviones* are here, summer has come", say the pensioners on the benches in Main Street, without even thinking of the long and arduous journey these birds – incredible flying machines that they are - have just completed.

The passage of passerines and other small birds like Hoopoes is perhaps not as visually dramatic as that of large birds of prey. That is probably why they have fewer followers. But they fascinate me and many of my ornithologist colleagues in Gibraltar too.

On the Rock, as there is a limited variety of nesting birds, any bird of passage is easily

spotted. If you see a Spotted Flycatcher, then it's a migrant. And the most fascinating thing is that if you watch carefully you can potentially see almost any European migrant bird and even some from further afield. That is why, especially as regards passerines, most of what is known and published about their migration in the area comes from our observations on the Rock.

As Gibraltar is only small, the birds tend to gather in certain small areas of suitable habitat well known to us, such as the military area of Windmill Hill Flats, where, when conditions are ideal, migrant birds abound.

With a combination of a good nocturnal passage, a light easterly wind and cloudy or misty weather, perhaps with a little drizzle, you can see birds absolutely all over the place. Flying by moonlight or guided by the stars but suddenly plunged into thick cloud, the birds will take refuge on the ground. Those that have begun to cross the Strait might turn back and others, hit by bad weather to the east or over the Bay and perhaps attracted by the flickering lights of the city of Gibraltar, head for the Rock.

Windmill Hill Flats, a military zone covering about 13 hectares, is an open habitat with some scrubland. Formerly at sea-level and shaped by prehistoric waves it is now about 100 metres above Europa Point, very near to the southern end of the Rock. So it is an ideal resting place for those birds which have just arrived from Africa during prenuptial migration and offers a last opportunity for rest to those who are crossing the Strait after the breeding season. Under the right meteorological conditions the place fills up with birds. I remember one year following prevailing southeast winds at the end of April 1984 when Windmill Hill became a meeting place for 37 species of migrant birds. During that spring 68 species of migrants were seen here, not including raptors or other species like swallows, house martins and swifts which are constantly flying overhead.

For me there are two times of the year which I most look forward to as regards migrating passerines. A drizzly morning in April brings back fond childhood memories. Quails, Hoopoes, Doves, Warblers, Pipits and Larks are usually associated with these memories. It was one such day, a 10th of April, when I saw a Tristram's Warbler, *Sylvia deserticola*, a species hardly ever seen in Europe. It was also in spring, this time in May and also on Windmill Hill, that I observed another of my "best birds", a male Bobolink (*Dolychonyx oryzivorus*) with nuptial plumage, far away from its meadows in the north of the United States and Canada and with only a female corn bunting to sing to!

But I prefer the autumn. Those September days when the fresher mornings bring a hint of the autumn soon to come, the westerlies bringing Honey Buzzards and a few clouds and the mountains in Morocco now standing out more clearly as the summer haze struggles to hang on. When already I can start to picture how green the countryside can be. It's the time of year when birds are dispersing after nesting and migration is starting in earnest. In Gibraltar those of us aware of nature of these changesstart hearing the call of the Iberian Chiffchaff, and see Short-toed Tree Creepers, Nightingales, young Shrikes, Golden Orioles and even the gaudy Kingfisher on the seashore or in the ponds of the Botanic Garden.

On autumn nights when low cloud embraces the cliffs of the ancient Rock, when the pines are shrouded in mist, you can drive along the upper roads of the Rock where you are sure to see

Opposite

The Olivaceous Warbler finds food and protection in the scrubland of the Upper Rock Nature Reserve.

The Wren is one of many
passerines that fly over the
Rock, and only pause there
on very windy days.

Scops Owls on fence posts , and both Grey and Red-necked Nightjars resting on the ground.

We are very lucky in Gibraltar to have an active ringing group led by Richard Banham and Charlie Pérez, who, with the help of other ringers from abroad, mist net every day during the migration seasons, ringing and then releasing them to continue their long journeys.

Ringing and observation over the years have shown that the majority of passerines that migrate at the beginning of autumn and come back later in spring are trans-Saharan migrants – those that winter in tropical Africa, south of the Sahara. So the Tree Pipits, Nightingales, Redstarts, Whinchats, and Orphean, Melodious and Spectacled Warblers almost always appear before the Mediterranean wintering species like the Meadow Pipits, Black Redstarts, Stonechats or Blackcaps.

Between 1991 and 2003 about 30,000 birds were ringed in Gibraltar. The most captured species was the Blackcap, followed by the Robin, then, curiously, the Black Redstart, with the Sardinian Warbler in fourth place.

The Black Redstart is easily spotted in Gibraltar in winter, both on rocky areas and cliffs and in the city. Large numbers of Black Resdarts occur also during the autumn passage in October and especially November. It is interesting to note that the subspecies which winters in the area of the Strait is *Phoenicurus ochruros gibraltariensis*, different to the one nesting in the mountains of Málaga and Cádiz, which is *P.o. aterrimus*.

The majority of the most commonly ringed birds, like the Black Redstart, are winter visitors to our area, what I like to call our Christmas assortment.

I always get a kick in the autumn when I first hear the typical call of the Robin or see the rusty flash of the Black Redstart's tail. At about the same time, the Chiffchaffs arrive, the population of Blackcaps increases by hundreds and you start seeing Stonechats that'll be around for at least several months.

A typical Gibraltar Christmas card would show a Chiffchaff perching on an *Aloe arborescens* flower drinking its sweet nectar. This is a very common sight in gardens on the Rock, not just confined to Chiffchaffs as Sardinian Warblers and Blackcaps also feed on the flowers. Sometimes, after bad weather when there isn't much food available, you can see many of these birds on the aloes, some of which defend their plants tooth and nail!.

Another Gibraltar typical wintering species is the Crag Martin, a bird of rocky areas whose preferred habitat stretches over a large part of the Rock. In the past a few pairs have nested, but now it just occurs on passage and in the winter. Their roosting places are in caves and tunnels, especially the large caverns near to where the Shags nest on the south-east side of the Rock. Ringing has shown that these martins come from the Alps, the Pyrenees and the Balearics. In the 1980s there were up to 2,000 birds in some roosts but these have decreased to a few hundred at the most. Perhaps some populations that used to visit us have disappeared or warmer climate may mean they don't migrate as much; or they may have found other roosting places in the area.

I hope they keep on coming. The arrival of these flocks in the evening, when they drop vertically keeping close to the cliff in order to avoid the talons of the ever watchful falcon, is one of the lesser known ornithological sights of the area, but one which I hope never disappears.

A very special wintering bird for us is the Alpine Accentor. Although at this time of year

This Black Redstart is a European
subspecies that winters on the Rock.

it appears in other rocky places in the area, it is only in Gibraltar where you can be guaranteed to see them, either singly or in pairs. To observe them you have to go to the summit of the Rock in the morning in late December or early January and patiently scan open areas of rock with scant vegetation.

When March comes, everything changes. The first Hoopoes, Swallows, Black Kites and some other birds have begun to arrive from Africa, while Robins, Pipits, Chiffchaffs, Black Redstarts and the Crag Martins have started to leave.

Sea birds

Sea birds have their own migratory cycle. We can pick it up in August when many of these species pass. In the evenings from the Europa Point observatory, especially when the winds

At the Punta Europa lighthouse it is easy to see *Audouin's* Gull flying just a few metres away from the bird watchers.

come from the south west, there are many interesting species to be seen.

One of the most common is Audouin's Gull (*Larus* audouinii) which can be seen flying west towards the Atlantic at this time. Adults, subadults and juveniles pass very close to observers at the Point and there are probably better views to be had from there than from anywhere else on the Iberian Peninsula.

Other gulls frequently seen in August, as well as in July and September, are Mediterranean Gulls (*Larus melanocephalus*), Black-headed Gulls (*Larus ridibundus*) and Lesser Black-backed Gulls (*Larus fuscus*). Other species passing through are Common terns (*Sterna hirundo*), Black Terns *(Chlidonias nigra)*, and Little Terns *(Sterna albifrons).* Less frequently seen are Caspian Terns (*Sterna caspia*) and Gull-billed Terns (*Gelochelidon nilotica*) and, on rare occasions, Royal Terns (*Sterna maxima*). All three skuas can be seen at this time, with the Great Skua (*Catharacta skua*) being the most common, but Arctic Skuas (*Stercorarius parasitius*) and more occasionally Pomarine Skuas (*Stercorarius pomarinus*) are both regular. Great Skuas stay over the winter and can be seen harassing the gulls throughout this season.

Avid sea bird watchers are kept busy in summer and autumn. Albert Yome, Paul Acolina, Harry Van Gils, Andrew Fortuna and Roger Rutherford as well as other colleagues already mentioned have spent hours and hours over the years scanning the blue and grey horizons, on the lookout for any sea bird close enough to be observed.

The passage of sea birds continues in October, which is definitely the best month to see a Lesser Crested Tern (*Sterna bengalensis*), two or three of which appear every year, possibly coming from one of the few colonies on the North African coast.

Gannets (*Morus bassana*) can be observed in the waters of the Strait throughout the year. In the summer there may be a few sub-adults in the area, but in August numbers start increasing and by September and October considerable numbers are passing, at first mainly juveniles, followed by adults which are the more frequent in winter.

One of the typical sights at sea in winter is on days when levante-driven storms force numbers of Gannets to take shelter in the Bay, when they can be seen diving for fish even inside the very port of Gibraltar. By this time several Sandwich Terns (*Sterna sandvicensis*) have already taken up winter residence, flying along the shore in search of small fish.

In some winters the easterly storms bring in numbrs of Little Gulls (*Larus minutes*), and I remember one December seeing hundreds of these attractive little seabirds all around the shores of the Rock.

As winter nears its end, in February, Cory's Shearwaters (*Calonectris diomedea*), on their way back from the Atlantic, also react to the wind by seeking the shelter of the Rock. When they get to Europa Point they have no option but to go out to

From left to right: Keith Bensusan, Zoë Cortés and John Cortés watching seabirds at the GONHS observatory at Europa Point.

the open sea and so pass close to the observatory from where you have a clear view of the complete flow, with thousands of birds passing hourly. In fewer numbers but also present in the Strait, on passage in winter and sometimes in the breeding season, are the Balearic Shearwaters (*Puffinus mauretanicus*). The closely related Levantine Shearwaters (*Puffinus yelkouan*) are seen more frequently in summer when, up to about fifteen years ago thousands would come to the entrance of the Bay to complete their moult. Sadly, this sight is no longer to be seen either in Gibraltar nor at the entrance to the Bay of Tangier, another place where these birds used to converge.

A few weeks after the Cory's shearwaters it is the Puffins (*Fratercula arctica*) and the Razorbills (*Alca torda*) which begin their migration, leaving the Mediterranean to head for their breeding colonies in the British Isles and other northern shores. From a distance they look like little more than small columns of dots flying rapidly low over the sea, especially in April and in the evening. The reason why they fly in the evening is not clear although there are observations in the English Channel indicating that they mainly fly in the morning. Is it at all possible that they make this extraordinary journey flying round the south west coast of Europe during the course of one, long, spring night?.

Because the reflection of the morning sun on the sea makes watching difficult, the best time to observe passage past Europa Point is in the afternoon. It is an exciting possibility to observe almost any species of sea bird usually present in the area, and perhaps one or two that usually aren't. And, of course, with the added value of the likelihood of spotting migrant raptors too, birding at Europa Point can be a very pleasant and enjoyable activity.

You can watch Gannets all year round from the observatory that GONHS has near the Punta Europa lighthouse, called *la farola* by locals.

The observatory at Europa Point, which belongs to the telecommunications company Gib Telecom, which makes it available to the GONHS, is the spot from which you can observe most of the very few shorebirds which visit Gibraltar. Some can be seen flying very low over the sea round the southernmost tip of the Rock. Sometimes one or two stop to rest or feed on tiny invertebrates on the rocks by the sea at the foot of the cliff. The most frequent are the Whimbrel (*Numenius phaeopus*), Oystercatcher (*Haematopus ostralegus*) and the Turnstone (*Arenarias interpres*) which, just like the Common Sandpiper (*Tringa hypoleucus*) sometimes occurs through the winter. The lack of marshland, large beaches and estuaries means Gibraltar is lacking in shorebirds and other wetland species, although many species, including Greater Flamingos, herons and duck, can sometimes be seen flying by. Some species which use the seashore as well as fresh water are seen more often, especially the Little Egret, the Grey Heron and even the Kingfisher, which are fairly regular outside the breeding season.

Gibraltar currently provides nesting habitat for two species of sea bird, the Shag or Cormorant (*Phalacrocorax aristotelis*) and the Yellow-legged Gull (*Larus cachinnans*).

During the 1970s and 1980s there was a huge increase in the population of Yellow-leg-

Puffins very near the coast after heavy levante storms.

ged Gulls and the military authorities on the Rock, responsible for the airport, began to take action. This initially involved destroying nests and eggs or poisoning the nesting birds. However, the population kept increasing until 1999 when the GONHS took over and started culling the adults with air rifles.

The population in 1991 was an estimated 30,000 birds at the end of the breeding season and including the young of that year. After the culling began this number was down to about 20,000 in 2002.

It was in 1981 that a nest was seen for the first time on a roof in the south of Gibraltar in Rosia Bay. Since then the number of nests on buildings has risen yearly. A rapidly growing bird population needs more nesting space and these gulls have now taken to nesting on the ground, in the scrub and even in trees, as well as buildings; and they have now colonized nearby towns such as La Línea and Algeciras. This close contact between gulls and the public has not gone down well especially because these gulls have no consideration for humans and make a persistent racket all through the night. Their faeces are also a potential health hazard, but this has not yet been confirmed.

Activity in the breeding areas is most intense in May when there are growing chicks in the nest. At that time the adults fly back and forth between the nest and the feeding areas. There is an average of two chicks per nest and they are fed almost exclusively with fresh seafood, despite the fact that large numbers of adults visit all the landfill sites in the area to feed on rubbish.

Large gulls have a great ability to adapt their diet and I have seen them feeding on rats and even catching Swifts in flight. They also eat fruit from trees and bushes, preferring red or orange

Yellow-legged Gulls will aggressively attack any migrant bird that strays into its breeding territory.

fruits like those of *Osyris quadripartita*. Because of this attraction to brightly coloured fruits they take fruit from the dragon tree (*Dracaena draco ajgal*) a Macronesian and African tree of the Moroccan subspecies found in the Botanic Garden in Gibraltar. Seeds of the dragon tree have been transported in the faeces of the gulls allowing this exotic tree to establish itself as part of the flora of Gibraltar. It is now naturalised on cliffs and slopes all over the territory.

The gulls are present in Gibraltar all year round, but in the months of August, September and the first half of October only a few hundred stay close to the Rock. Ringing has shown that they fly off to various parts: in the Mediterranean they have appeared from Málaga to Almería, in the Atlantic as far as Cape St. Vincent, and one was found in autumn not far from Paris. One unconfirmed report of several yellow-legged gulls with blue rings and three white letters – the ones used on the Rock – places them in a port in Mauritania. In October, however, adults and some sub-adults return to the Rock to take up nest sites. Around February they start building nests where they will lay from one to three eggs.

In June and July the dark-coloured young leave the nest and some go straight to the water where they dot the sea just a few metres from the beach in Gibraltar and La Línea. But others fall onto roads, car parks etc., and are often run over. While the fledglings are still in the nest the adults try to scare off potential enemies (including humans) by diving at their heads. Direct impacts are few and far between.

They are acutely aggressive towards large raptors, harassing vultures and Short-toed Eagles in particular. The Peregrine Falcons also have chicks at this time and often join forces with the gulls to chase the raptors away.

Quite frequently the raptors, heavy and tired, seek refuge on the ground while still under attack. Other times they fall into the sea, and unless they are rescued by a passing boat, they are drowned – a sad end to a long journey.

Whenever possible these birds are picked up and taken to the GONHS rehabilitation centre where Vincent Robba and Stanley Olivero have a long list of successes in the recovery and re-release of raptors. In 2002 two Short-toed Eagles, a male and a female, were released with a satellite transmitter and both survived. One sent signals until the end of the summer from the south of Extremadura, whereas the other stayed near Gibraltar all summer, flew south to Morocco in winter and then returned the following spring. The bird of prey rehabilitation unit has rescued and released many birds since its inception. These have included a female Bonelli's Eagle and, as we have seen already, a female Eagle Owl. Lesser Kestrels and Peregrine Falcons have been successfully bred and raised in captivity and there is no lack of new projects for rehabilitation and reintroduction of species.

It is fitting that, on this small rocky, coastal territory, which advertises its nature on postage stamps and coins, there is also such a direct contribution to the conservation of the natural heritage shared by all the inhabitants of the Strait.

In general Gibraltar can be considered to be a focal point for migration and natural history in an area where it is still found in such abundance. Much is shared with the rest of the Strait, but it is also very different. It is small, but channels a large proportion of the birds which pass through the Strait and observations from the Rock have also contributed a great deal to what we know about the phenomenon of migration.

Opposite

Yellow-legged Gulls nest on all the buildings in the coastal towns of the Campo de Gibraltar.

It was an angled formation of birds
aiming for
that latitude of iron and snow
advancing
relentlessly
on a straight-lined course:
the all-consuming rectitude
of a visible arrow,
numbers in the sky going forth
to multiply, in lines formed
by the imperative of love
and geometry.

Pablo Neruda
Migration, 1971

Migration in general and special characteristics of the Strait

Migration in the animal world is the seasonal movement undertaken by animals of the same species to search for food or to reproduce in a different habitat from their current one, to which they later return. It is a kind of natural transhumance organized by the animals themselves, without outside intervention. Although it never ceases to amaze, it is a phenomenon identical to that of man, who, from his origins until this very day has made human migration one of the characteristics of the species.

Nearly everyone has heard of the migration of gnus in the Serengeti between Tanzania and Kenya; of the monarch butterfly which winters in Mexico; of the salmon swimming upstream in virgin rivers, where a crowd of impatient bears awaits them; of the slippery eels, chubby tuna fish, striped zebra, geese, storks etc, an endless number of animal species which migrate to a different habitat, only to walk, swim or fly back to their starting point a few months later.

What is it that triggers an animal's urge to change its habitat? How do they know that their time in a given territory is up? When they decide to leave, do they start spontaneously or do they feel it coming days before? Some such questions have clear answers but others remain either unanswered or there are only theories about them. But nearly all authors and scholars agree that migration, as a general rule, is a nutritional or sexual problem; animals migrate when they *know* that historically there comes a time when food resources are lacking or when they have to breed or give birth. One of the odd features of migrations is that, though they are usually two-way, this is not always the case. One example is the salmon, which goes back to the river where it was born, reproduces and then

dies; or the Griffon Vulture (*Gyps fulvus*), whose young migrate the first year but are sedentary when they return.

The gnus leave because their pastures are depleted and the salmon because they have to reproduce. Generally these migratory movements are preceded by other pre-migratory dispersive movements. Animals do not leave their territories in a mad rush, but do so by dispersing first to neighbouring territories. Gradually they get together with other individuals, which have arrived in the area by instinctive urges, until they form a flock or herd, depending on the animal, and decide to begin migration.

Invertebrates like the monarch butterfly, African locust or South American legionnaire ant all migrate, as do fish such as tuna, eels, sturgeon, trout and salmon. Amphibians and reptiles are vertebrates which tend to migrate less because they cannot overcome high altitudes, formidable barriers impossible to cross as they cannot withstand the cold temperatures. (An exception perhaps is the sea turtle, but these great swimmers are in their element, the sea, which is not a natural barrier.) So, countries like Morocco have, in a short longitudinal area a wide variety of reptile species; the Atlas mountains isolate species from one another and local movement is very difficult, not to mention migration. The maximum exponents of migration in the animal world are birds: geese, cranes, storks, etc., forming armies of more or less detectable migrants and most of the public at large is aware they are affected by this phenomenon. Finally, we have seen many times on television the vicissitudes of zebra and gnus crossing the river Mara, the natural frontier between Kenya and Tanzania, where monstrous crocodiles are awaiting them with voracious appetites. Polar

The Grey Seal *(Halichaerus grypus)* is an example of animal nomadism in sub-Arctic and temperate waters of the Northern Hemisphere.

bears, orcas, reindeer, whales, elephants and so on and so on are down on the list of migrating mammals until we get to man himself, a species of which there are fewer "classical" nomads and more and more immigrants, due to the economic, social and cultural changes which have taken place over the last few years.

It is a strange fact that, while some species abandon their territories in the Iberian Peninsula for lack of resources (raptors, storks etc.) others arrive here (although not in the same habitat) looking for food (cranes, geese, waders etc.) What for some species is a breeding area, for others is a wintering area. If we rely on the human premise that "you belong to where you are born" then Bee-eaters, Rollers, Bonelli's and Booted Eagles are European species. However, if plumage is anything to go by, then these are typical African birds.

For centuries man has wondered why birds migrate. There are several factors brought to bear on whether a bird decides to migrate or not. Photoperiodism – the period of exposure to daylight – affects the pituitary gland, in such a way as to liberate the hormones controlling the biological clock, which stimulates the bird's desire to leave its territory and, along with other hormones, act on its fat metabolism and moult. This process is fixed for each species in each geographical location. That is why migration takes place before food reserves start diminishing; it is a biological safety mechanism which prevents resources from running out. But, as we shall see later, there are species in which photoperiodism is not the only decisive factor. Some storks, for example, do not migrate and spend winter in the Peninsula. The food factor is logically very important, especially when cut off drastically. If an early snowfall finishes off all the insects in an

area, insect-eating birds will have to start migrating or they will perish.

Once their stay in the nest is over, juvenile birds make dispersive movements, which consist of erratic movements taking them a certain distance away from the breeding area. Sometimes they fly many kilometres in the opposite direction to that which they will later follow during migration. This behaviour suggests they do not yet feel the urge to migrate and are curious to explore other territories in no matter what direction.

In the Iberian Peninsula this post-generational dispersion is typical of young Booted Eagles. Movements north-eastwards are recorded every autumn in the opposite direction to migration, which is south or south-eastwards. In the migration forum on pernis@eListas.net these movements by Booted Eagles from north-eastern Spain over to France have been avidly debated and it has even been suggested that it could be a migration "going the wrong way", i.e. they cross the Pyrenees, go across south-east France and into northern Italy, then carry on towards the Strait of Messina.

Bird migration can be classified according to the length of the journey undertaken. Those involved in long distance migration usually live in the northern hemisphere, in lands quite near the North Pole with relatively warm summers and plenty of food but the reverse situation in winter. An example is that oft-quoted, long distance migrant, the Arctic Tern (*Sterna paradisea*). One which was ringed on the coast of Russia's White Sea was monitored in Fremantle, Western Australia having flown 22,530 km. Another example of long distance migration, this time of a relatively large bird, is that of the Osprey (*Pandion haliaetus*). In autumn 2003 some Finnish scien-

Flock of White Storks on postnuptial migration flying past a ship going through the Strait.

tists fixed a radio transmitter to a male Osprey and followed its signal from 20th September to 16th November when it arrived in South Africa, having covered 12,500 km, crossing the eastern Mediterranean and flying the rest of the way overland down through Africa. Its movements on the return journey were also monitored as the radio battery did not give out. Funnily enough, in contrast to its mate, the female wintered in Equatorial Guinea.

Short distance migration usually involves birds which live in territories north of the Mediterranean and fly to the coast where they find a somewhat milder climate. Small birds like the Chaffinch (*Fringilla coelebs*), Goldfinch (*Carduelis carduelis*), Robin (*Erithacus rubicula*), Song Thrush (*Turdus philomelos*) or Dunnock (*Prunilla modularis*) undertake these short journeys along with Lapwings (*Vanellus vanellus*), Pigeons (*Columba sp.*), Cranes (*Grus grus*) and other larger

A spectacular flock of White Storks not far from Tarifa on a day of strong poniente winds.

species. In the southern hemisphere where the continents are farther away from the South Pole, seasonal changes are less pronounced and birds do not feel the urge to migrate.

The age and sex of some migrant birds may mean changes in the way they migrate. As regards age there are species, such as Griffon Vultures, in which only the young of that year migrate, or, in the case of White Storks and many European raptors, postnuptial migration is started by juveniles, unaccompanied by the adults. Prenuptial migration in this group of birds is started by the adults and there are even species which do not return until they are mature, such as the White Stork and, possibly, the Griffon Vulture. Among sea birds we have the example of young Gannets (*Morus bassanus*) born in Scotland and migrating to the west coast of Africa, while the adults disperse along the English Channel and the Bay of Biscay. As for their gender, there are species such as the Chaffinch where the females and the year's young migrate and the males remain in the breeding area. In prenuptial migration the male White Stork, Nightingale and Harrier will arrive first at the breeding site, probably to try and get the best nesting places.

Bird migration

Birds in flight move mainly in two ways: by flapping their wings or soaring. There is a third way, halfway between these, in which the bird alternates soaring with wing flapping.

Flapping is typical of birds with great flying capacity, such as passerines and falcons, although probably the most famous example is the Hummingbird, *Heliactin comuta*, from South America which flaps its wings 90 times per second. This form of flying uses up considerable muscular energy and therefore a bird must

keep its fat reserves up (or at least administer them conscientiously to keep going). One way of making progress at great speed and reducing energy consumption is to alternate flapping with prolonged soaring. Many birds do this, including the Common Swift (*Apus apus*), which manages to cover long distances very fast and uses up less energy than passerines. The advantage lies in the fact that they need less time to refuel and therefore less time to migrate, which is when many birds perish due to bad weather, predators, lack of food etc. Migration is indeed an adverse situation and the sooner it is completed the better it is for the bird. The Swift, while flapping and soaring, feeds in mid-flight, an added virtue.

Arrival of a flock of Griffon Vultures on the north coast of the Strait, forced to drift by the strong poniente. In the absence of drift they would arrive directly in front of the observer.

The Glittering-throated Emerald Hummingbird (*Amazilia fimbriata*), is an example of a wing-beating bird that must consume daily the equivalent of its weight in nectar from flowers.

Soaring means making progress or staying up in the air without moving any wings and taking maximum advantage of the energy in the atmosphere. Pure soaring birds don't exist, however, because when rising or landing all birds beat their wings, likewise when making sharp turns or picking up speed. So, soaring birds are those which devote the greater part of their time in the air to soaring. Soaring is characteristic of large birds which would otherwise consume a great deal of energy when flying and need to eat almost constantly.

One of the advantages obtained from soaring, apart from saving energy, is that birds do not need to fly with large reserves of fat, as is the case of flapping birds, and can therefore travel long distances with little *fuel*. This does not mean that soaring birds do not store up fat beforehand, but store less. Nevertheless, it is still amazing that Honey Buzzards do not feed at all during their long journeys in autumn and spring, although it is also true that it is the raptor which accumulates the largest reserves of fat in its body for migration, possibly as a precaution against its not feeding during that time. I only know of two occasions when *pernis* (as Honey Buzzards are colloquially known in Spain) were observed feeding during migration: one on prenuptial that I was fortunate to witness myself and another on postnuptial that was seen and photographed by Joaquín Mazón, who is involved in the Migres programme. This is a great exception to the normal behaviour of this migrating species.

Something inherent in soaring birds is their use of the air currents called thermals, which are rising columns of warm air caused by atmospheric convection. The air near the ground is heated by the sun, thereby losing density and causing it to rise vertically. This mass of air is then occupied in turn by another, but colder, descending mass of air. These thermals rise up to 300 or 400 m in our part of the world, but in warmer parts of the earth like the tropics they may go up to 4,000 metres.

Soaring birds travel along using thermals and alternating them with gliding. This is called termed thermal soaring. They convert potential energy (altitude) into kinetic energy (soaring) and the bird's use of energy is minimal. A typical situation would be as follows: the soaring bird waits until the morning sun is warm and thermals start forming; at the appropriate time it flies, beating its wings, until it locates a thermal and then, by *circlings* or corkscrew-like movements within the thermal, it starts to rise; when it has gained sufficient height, it leaves the thermal and, going slightly downwards if there is no wind, it starts to glide until it finds the next thermal which will take it back up to the previous height, and so on. Once it has gained the height it needs and if the wind is strong, the bird can keep going without flapping its wings. Depending on the time and place whole trains of thermals can be formed – by slopes facing the sun, rocks etc. – into what ornithologists call corridors, which are often used by groups or flocks of migrants. Moderate to strong winds blow all the time in the Strait and move the thermals. This movement is seized on by the soaring birds, which corkscrew their way up the thermal and go in the direction the wind is blowing at the time. Thermals do not usually form over the sea and large expanses of water, like lakes and wide rivers, but, if they do, they are weak with little upward thrust. That is why soaring birds try to avoid crossing the sea and choose straits instead.

As we have said, the turns inside the thermals made by the soaring birds are called *cir-*

Opposite

A thermal forming in the Sahara desert. Observe the air rising and dragging sand with it.

clings. These, like an ascending corkscrew, push the bird vertically upwards and it must decide when to leave one thermal for another. Within the same thermal one can observe birds circling clockwise and others anti-clockwise, all achieving the same results in elevation, although when the thermals have a small radius they have to circle more quickly than in larger thermals. In order to rise more quickly a bird must circle as near as possible to the air current which will take it upwards at the speed of the thermal.

Another form of soaring, termed dynamic soaring, is used by sea birds and consists of using the weak air currents generated at sea when the air hits the waves. The bird flies into the wind and gains height. When it reaches a neutral point it turns and starts to descend, with a tail-wind, until it reaches a point where once again it begins to rise and the cycle begins once more.

Other kinds of currents used by soaring birds to travel are the slope currents, formed by winds colliding violently with the sides of the *sierras* and mountains and causing great masses of air to rise to considerable heights. Soaring birds use these columns of rising air to move upwards without wasting energy and get over high ground which would otherwise have meant an additional effort, through wing flapping, in their migration. Any saving of energy, no matter how small, is very important for the survival of migrant birds, especially when they have covered many tiring kilometres.

Converging air currents are yet another type of very useful current for soaring birds and they are formed at the meeting point of two masses of air coming from opposite directions. This phenomenon is frequent in the Strait, both at sea and inland. It is quite common to see a ship whose smoke is going eastwards and another, several kilometres distant with the smoke from its chimney going west; as they approach each other, the smoke trails become more and more vertical until finally the smoke from each ship is going in the opposite direction to that before they met. These converging currents, not easily detectable by the casual observer, help the migrant bird to rise and progress in the middle of the Strait with hardly any effort. They are responsible for the migratory corridors, which those of us who spend hours watching migration in the Strait have seen many times without finding a reasonable explanation for them.

Flight altitude depends on the migrating species, the weather, relief, time of day etc. With regard to the altitude at which migrants fly, these may be altitudinal migrants which migrate by flying at high altitudes, either because their habitat is in high mountains or because they have to fly over them to arrive at their destinations. One good example of an altitudinal migrant is the Bar-headed Goose (*Anser indicus*) which flies over the Himalayan mountain range in Nepal, nearly 9,000 m high, to winter in northern India. But the highest altitude on record for a bird is the 11,277 m reached by a Rüppell's Vulture (*Gyps Ruepellii*) which collided with an aircraft in 1973. Nearly all nocturnal migrants fly at greater altitudes than daytime migrants to avoid mountain ranges. The majority of low flying migrants are soaring birds which need thermals to go higher and then glide to conserve maximum energy during migration. Storks and raptors do this when crossing the Strait.

Cruising speed is defined as the ratio between the total distance covered and the total amount of time taken. In soaring birds the time taken involves gliding and circlings, which makes them much slower than the flapping birds whose time

Lesser Black-backed Gull skimming
masterfully over the waves of a heavy swell.

is equivalent to net progress. This speed is not constant because it includes stops for eating and resting, which during migration, and depending on the species, may be considerable. Cruising speed depends on various factors, but primarily the species in question, as there are slow flying birds like Pipits which fly at 40-50 km/h and others like the Peregrine Falcon whose cruising speed is 100 km/h. Logically these speeds depend very much on the prevailing wind, which, if it's a strong headwind, will reduce their speed and, if it's a tailwind, will increase it.

The timing of migration also varies depending on the species involved. Passerines tend to fly at night as it makes them less vulnerable to predators, whereas raptors usually fly by day to travel with the added advantage of thermals and because they do not fear predatory attacks. According to some recent studies in Germany

José Ramón Benítez and Manuel de la Riva attaching a radio transmitter to an Egyptian Vulture chick for satellite tracking.

many birds migrate at night in order to benefit from lower temperatures, which are less energy-sapping, especially in those which have to fly over deserts, where daytime temperatures would make flying almost impossible. For this reason soaring birds start their migratory movements well into the morning, when the sun is heating up the earth, and they finish at dusk when thermals stop forming. In the middle of warm days soaring birds may fly at altitudes of over 800m. This makes them invisible to observers, who are inclined to think that migration has ceased during the hottest part of the day.

Orientation in migratory movements differs according to species, time of day and date. For many birds, their destination is recorded in their genes and they don't need others to migrate with. It is instinctive behaviour with no learning process involved. It is thought, therefore, that these migrants are governed by a genetic component, whereas other species have memorized the routes to follow. An example is the adult Cuckoo (*Cuculus canorus*), which lays its eggs and just flies off; juvenile cuckoos have to migrate by themselves without being led. This also occurs with White Storks and Black Kites – the juveniles leave first and are followed later by the adults. However, it is not only important to inherit the migrating instinct but also to acquire, genetically, the annual cycle that triggers off the physiological changes necessary for them to successfully undertake the migratory journey.

How do birds locate the flyways that take and bring them back on their annual migrations? Today we know that different species use different systems. Some guide themselves by the position of the stars: experiments carried out with nocturnal migrants in planetariums have proved this to be so. Experiments with daytime

Migrating Black Kites. They cross at straits because they are unable to cover long distances by continuous wing flapping.

migrants have led to the conclusion that they are oriented by the position of the sun and the time of day, added to which there is the all-important reference of the terrain. Genetic impulse takes them towards a specific destination and, on the way, they memorize physical details of the terrain covered; on the return journey the genetic component is no longer so vital because they are guided by memory. A clear example of this is the case of a young Egyptian Vulture, which had been ringed in Andalusia in 2003 and mi-grated to Mauritania in a straight north-south line, without straying from the route. After wintering there it started on the return journey, but was blown off course by a strong sandstorm in the Sahara desert and ended up near the Atlantic coast. When the winds calmed, the bird flew back to the point at which its journey had been interrupted and continued in the south-north direction taking it back to Andalusia. This behaviour suggests that during the outward journey it followed a natural impulse recorded in its genes, memorized the route and, after being diverted by the sandstorm, it doubled back and followed the original route. (Benítez *et al.*, 2002, *Conservación y recuperación del alimoche en Andalucía*).

Using the magnetic field is another important way to navigate and experiments with pigeons by attaching a magnet to their heads have

Flock of Sparrows one evening at the former La Janda lagoon. During migration large numbers of birds stop off here to both eat and rest.

demonstrated this. It has even been shown that in the daytime they can orientate themselves by the sun and if it disappears behind clouds they use the earth's magnetic field. There are some migrants, like Bee-eaters (*Merops apiaster*), which migrate both by day and night, alternating visual navigation and use of the Earth's magnetic field. On hot days at the end of August in the town of Algeciras, when windows are opened at night to let the sea breeze in to cool down the houses with no air-conditioning, you can actually hear the musical sing-song of migrating Bee-eaters. I can assure readers that it gives me special satisfaction at home when I hear them passing, since it reminds me that migration doesn't stop, even at night!

Depending on the species, migrations can be undertaken by individuals, groups and flocks. Individual migrants are generally birds with a powerful wing beat, usually raptors; birds like Harriers with a large wing span take advantage of any air current to make progress, and a third group uses the thrust of the wind to glide and, in its absence, flap its wings, like the Osprey over both land and sea. These birds, falcons, harriers, ospreys and hobbys, among others, do not need thermals to travel and therefore do not aim for the straits because they can fly over the sea with ease, using their muscular strength and flight strategy.

Migratory groups are formed by recruiting members at bottlenecks which build up at, in our case, the Strait. Birds feel the urge to migrate and take off on their own, but on the way they come across others of their species and fly together. They do not form a flock in the strategic sense, although as they coincide with others on the same route, they may look like one, if rather spread out. The prime example in the Strait is the Short-toed Eagle and, on occasions, the Booted Eagle.

Finally, flocks are made up of individuals which like to fly in a compact group with others, having a common strategy and benefiting from one another's company. The flock makes the individual feel safer when faced with bad weather, difficult terrain and even feeding difficulties. Travelling in typical V- formation it has been proved that flying just behind others helps birds like ducks and geese save energy. In this way they avoid the air disturbances caused by the one in front and their vision is not impaired. White Storks fly in compact flocks and create vortices for the benefit of the whole group. However, flying in flocks is not always possible, e.g. in unfavourable weather conditions or when the strongest birds are separated from the weakest, which occurs with Griffon Vultures crossing the Strait. Passerines also group together in flocks when migrating and this may help to keep predators away.

As mentioned previously, the accumulation of fat before migrating is essential so that birds can travel long distances. This accumulation of fat reserves in the body is a reserve of energy which allows them to successfully compensate for the weakening of muscles involved in wing flapping and, in the case of soaring birds, keeping them spread. It can be compared to fuel which is burnt during a flight. Some birds reduce their stomach size in order to lose weight, accumulate more body fat and in so doing increase flight autonomy. Some warblers, which usually weigh about 17 grams, build up to 34 grams before they start to migrate. Although the accumulation of fat is common to all birds it is greater in flapping birds because they use up much more energy than soaring birds.

Migrating flock in V-formation; the birds save energy by flying in the slipstream of the one in front and avoiding turbulence.

The Swallow is a well-respected bird because, according to a Christian tradition, it took the thorns from Jesus' crown when on the Cross. It only nests in buildings in use.

Migration: origin and destination

Breeding and wintering areas may be hundreds or thousands of kilometres apart. There may be a greater abundance of food in the breeding areas, given that these manage to withstand trophic pressure, at least while the chicks are being fed. Large-size migrants like storks, raptors or cranes lay one clutch of eggs, whereas passerines lay several to make up for losses caused by migration, accidents and predators. In these breeding areas birds are territorial, i.e. they do their courtship displays, nest building and repairing, bring up their chicks and start migration from this point. Adults have no difficulty in returning to their territories after wintering in far off lands, because they have memorized the terrain, but their young, who have never made the return journey before, also come back to the same place and sometimes to the same nest. Although some birds, like Cranes, fly off to wintering areas together with their offspring, most birds do this independently.

On one occasion in the Czech Republic a male Black Stork (*Ciconia nigra*) and a juvenile left the nest and flew separately to the Strait of Gibraltar, while the female and two other juveniles migrated via Israel. From this one can deduce that the family unit doesn't exist and the only link between animals is at reproduction time or on the nest. After that there is no fidelity, unlike in humans. This has also been observed in Ospreys from Finland, as we commented earlier.

Topography, weather and migration are often inter-related. There are birds which migrate to areas occupied ancestrally by those of their species, such as the Lapwing, which in some years may or may not arrive at certain latitudes. Whether it does depends almost exclusively on the weather. If there are heavy snowfalls lapwings will stay in the southern part of their traditional wintering area, but if the weather is even colder than usual there, then they may fly even farther south and appear in areas where they have never been before. Adverse weather also affects the timing of migration, often reducing the number of participants or stopping it altogether, depending on species and conditions. Heavy storms do not prevent migrations but sometimes cause them to be diverted and slow down the cruising speed. Foggy weather, if very persistent, might influence migrants, especially near the Strait. In both pre- and postnuptial migration we have noted an absence of migrants during days or hours of thick fog. If it is low lying fog, then it proves to be no obstacle as birds can still see the terrain from a different perspective above; however, when fog

is quite high then it may stop soaring birds from passing. Sometimes over the Strait in Morocco you can see mists covering the 800 m high Jebel Musa Mountain, which together with the Rock of Gibraltar formed the two Pillars of Hercules on each side of the Strait.

Another perhaps even more important factor than the weather for many flying species is the topography. We have mentioned how soaring birds, like Griffon Vultures, storks and Short-toed Eagles dislike flying over mountain ranges and wide expanses of water, yet other migrants, like the Bar-headed Goose, take them in their stride and fly over summits rather than go on a detour.

The most globally important sites for observing migrant birds are directly related to the

Areas of passage of largest numbers of birds of prey :

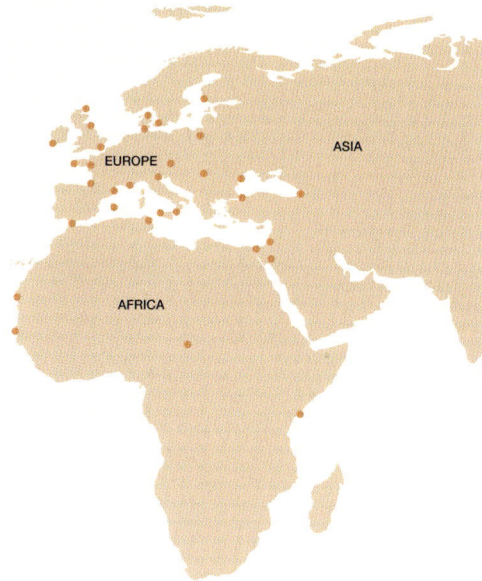

Soaring flock of White Storks. To save
energy they alternate soaring with wing
flapping and circling while advancing
across the Strait.

number of birds that can be seen at a certain time, although some ornithologists classify sites according to the variety of species observed and their importance from the human perspective. We are going to concentrate on those places where it is possible to see the greatest number of soaring birds, the ones which are most followed and appreciated by ornithologists.

The two areas where it is estimated that over a million raptors pass are Eilat in southern Israel (the confluence of birds from three continents: Africa, Asia and Europe) and the Panama Isthmus in Central America. In this ranking – obviously quite arbitrary – second place is occupied by the Strait of Gibraltar in the western Mediterranean and in the Bosphorus (Black Sea Strait) in the eastern. There are many more major areas where you can observe soaring birds, like the Pyrenees, Falsterbo (Sweden), the Messina Strait (Italy), Cap Bon (Tunisia), Kefar Kassem (Israel) and so on. Other good sites, although lower in the ranking, are Hawk Mountain in Pennsylvania, Goshute in Nevada and Djibouti in Northeast Africa. As far as different species are concerned, in Kefar Kassem 25 have been counted and in the Strait of Gibraltar 20. About 390,000 Honey Buzzards and 110,000 Lesser Spotted Eagles pass through Keffar Kassem; at Eilat more than 350,000 Honey Buzzards and 325,000 Common Buzzards; over the Bosphorus nearly 500,000 White Storks and in the Strait 120,000 White Storks, 120,000 Black Kites and 100,000 Honey Buzzards. These are quite spectacular figures for any avid birdwatcher.

Migratory phenology

Migratory phenology is not the same for every species – nor is it for every author – and, even within a species, there may be changes in any one year in one or both of the seasonal migrations. Some authors divide migrations into spring and autumnal, but birds do pass in winter and in the middle of summer too; others refer to pre- and postnuptial migration which would include these anomalies; and still more put migration in two periods: from January to June and from July to December. But putting this controversy to one side we can safely say that there is a migration prior to reproduction and another following it, whatever the name we label them with. For the Strait we will adopt the terms pre- and postnuptial migration, and perhaps autumnal and spring too, although they are more imprecise.

An outstanding feature of prenuptial migration is that it is spread over a longer period, whereas in postnuptial the passage is more concentrated – they are in a greater hurry to leave! This fact is known to ornithologists and rela-

A group of ornithologists from Sevilla at the Algarrobo observatory during the September passage.

tively few turn up to watch prenuptial migration because there is less chance to see a lot of birds in only a few days.

As far as the bird watcher is concerned, prenuptial migration begins in the Strait in mid-February, with the arrival of Kites and Booted Eagles, and ends in early June. Postnuptial migration begins in the second half of July and ends in the first half of October. These dates, of course, have nothing to do with bird phenology; they are just the time of year when most ornithologists from further afield come down here to bird watch.

Tracking of migrants in the Strait

The capacity of humans to watch the movement of birds is in fact quite limited. There are detectable migratory movements, from a subjective point of view, either with the naked eye or with additional optical equipment (binoculars and telescopes). Every study carried out in the Strait up to now has used this method (spotting and physical counting by observers placed in the area), but new methods are appearing, thanks to modern technological developments, and changing some aspects of studies. In this book we are going to have a look at how these changes affect the Strait and adjacent areas. Undetectable migratory movements are those in which migrants pass overhead without being detected for a variety of reasons, the main ones being the birds' flight altitude, weather and light conditions and, to a lesser extent, the colouring of birds. If a census of migrants is being undertaken then these undetected migratory movements reduce the final total to a minimum. One of the most important tasks of counting teams is to cut down the number of these movements, so that the census recording is as accurate as possible.

Luis Barrios, one of the coordinators of the Migres Programme, told me that with the aid of night vision binoculars he had been able to watch a flock of Honey Buzzards arriving at the European coast at 21:00 hours in May. This is a hitherto unknown and quite amazing occurrence, considering the pattern we are used to in this species. Therefore, we must remember that whenever we refer to a specific number of migrants crossing the Strait we are always referring to the minimum number detected.

The traditional counting methodology for daytime migrants has been used in two post-nuptial migration campaigns, using quite a number of observers. They were supervised by Francisco Bernis and Manuel Fernández Cruz, Professor in the Vertebrates Dept. of the Complutense University of Madrid. Although it is true that direct observation is not the most perfect method of precision counting, if the field work is continuous then errors are reduced to a minimum- this is what the Migres Programme terms *intensive effort*. But let's consider what teams in the field actually require if they wish to do daytime counting on the European side of the Strait for migration in both directions. This method, with some fine-tuning, can be extrapolated to other regions: Choice of suitable observation sites, minimum number of observers, knowledge of wind patterns in the area, observers' knowledge of birds, supply of binoculars and telescopes, familiarity with local bird life, weather data recording equipment (compass, thermometer, anemometer...) and radio-communication systems.

The choice of suitable sites for observers is of vital importance because the streams of birds arriving from the hinterland of the Strait, on both shores, do so at certain points, depending on the

Swifts mate, sleep and eat in the air. Migrating flights don't take much out of these fast birds.

prevailing wind at the time of arrival. Therefore, it is important to place observers in areas nearest to which birds are leaving from, during postnuptial passage, and arriving at, if it is prenuptial.

If it is important in postnuptial passage to have at your disposable a minimum number of observers – and this applies to observatories too – it is even more so in prenuptial passage because birds arrive well spread out, fan-like, due to the effect of the strong winds on migrants, especially in mid-crossing. As winds are generally from the side in relation to the Europe – Africa axis, birds in passage are thus blown away at right angles to the line they try to follow during the short crossing. Stronger individuals or species are subject to less drift than others and consequently arrivals at the coast are spread over a wide area.

Knowledge of wind patterns in the area is important in order to be able to put observers in the right places, as the strength of winds in the Strait is such that it can make them drift to one side both when arriving and leaving and if you are at the wrong watch point then you may not be able to see the main flow of birds.

Observers, in our opinion, must have adequate knowledge of what they are doing when identifying individual migrants or occasional species and detecting flocks. It may well be that a flock of Honey Buzzards is mistaken for Black Kites when travelling at altitude, the sky is cloudy or the birds are quite a distance away. This may not affect the final total to any degree if the error is not repeated, but it can be decisive as far a certain species goes. The same thing would happen if a bird were identified as a Spotted Eagle (*Aquila clanga*), Osprey, Imperial Eagle etc, when it is really another quite different raptor. In such minority species errors, as a percentage, can be scandalous, because an individual may represent up to 50% or even 100% of the sightings of the species in the Strait. Furthermore, without adequate knowledge and guidance, during postnuptial passage, observers at a site may all be looking in one direction while birds may be passing unnoticed behind them and, as we have often seen, several thousand can be missed by the end of a single day. Besides that, in both passages flocks may "sneak through" (as we say in ornithologist slang) at altitude, without being spotted, because observers do not look towards the highest point in the sky.

A good supply of binoculars and telescopes is an obvious item so we will not dwell on this except to say that binoculars are the better of the two because with them flocks can be monitored better than with telescopes, which amplify more but have a narrower range. Another disadvantage of telescopes is that when a strong wind is blowing in the Strait it is practically impossible to work with them, but they are useful for observing individual birds and monitoring flocks from a distance.

Familiarity with local bird life is useful if we want to avoid repeated counting of the same individuals. To do so there are several methods which allow us to determine if a bird is local or not. Phenology can help the observer a great deal, because if you spot a flock of Griffon Vultures in August, they must be local birds since migration does not occur until October when they start arriving in the area. If a Common Buzzard is spotted repeatedly from a certain observatory, then it is logical to think it is a local bird because they are not usually seen in the Strait. It is also possible to tell the difference between locals and migrants by the way they fly and as an example let's consider a Short-toed Eagle hovering: these eagles don't hunt when migrating, except when

On a cold, January morning in 2005, a small group of Cranes flying low over La Janda.

Large flock of White Storks, mainly juveniles, attempting the Strait crossing at low altitude.

they are held up by strong winds, so therefore the one in question must be a local bird. On the eastern part of the European coast of the Strait some pairs of Kestrels nest in the cliffs and in the beacon towers and if you watch their movements carefully you can see they are not migrants: birds of passage fly, determinedly, perpendicular to the coast, whereas the local ones come and go, hover and fly low.

Instruments for recording weather data (compass, thermometer, anemometer) are essential in scientific work to be able to interrelate different parameters of potential importance in any project. With the compass we can determine the directional tendency of the migrants, with the thermometer the temperatures of the observatories and with the anemometer the wind strength.

A permanent radio – communication system enables all observatories to participate in the arrivals of streams of birds, both in pre- and post-nuptial migration and to detect flocks, which would otherwise remain unnoticed. Moreover, at any given time all watch points can report temperature or wind data together and compare them later on.

The counting method for daytime migrants has also been used in the Strait but, instead of the *continuous method* in the many observatories along the coast, for financial reasons the one used is the so-called constant effort method, and it basically consists of observing from only two points along the coast, independent of each other and with no radio link-up, and treating the data objectively with no interpretation of them on the part of the observers. All this forms part of the Migres Programme for the 2002 – 2005 period. It remains to be seen if the method is effective with only two observation posts and just counting the number of birds flying overhead.

No doubt there are many people who, unfamiliar with the way ornithology works, will smile at some of the figures of migrant birds recorded by observers. Such figures could be, say, 1021 or 2375 and more than once I have been asked how you can possibly count exactly 1021 migrating White Storks and why you get such an apparently arbitrary number. To convince the listener I have to explain that, along the coast there are five separate observation posts, from which the flocks of storks can be seen, each manned by 3 or 4 people. When a flock approaches the first post in a thermal the storks are not counted when circling but by each observer when moving on towards

the next post between one thermal and the next thermal. There are several counting methods. I count in twos, taking half the time I would need if I counted individual storks. Once each observer at a post has a tally, then an average number at that post is obtained: quite an accurate figure for the size of the flock. But that's not all. An average figure is taken at all the posts the flock has passed and the result is very accurate indeed.

The counting methods for nocturnal migrants are by using radar or by direct observation under a full moon. We will not be dealing with them as they are outside the scope of this book.

The counting method for individual migrants is visual, like the previous ones, but an individual is monitored instead of a flock. It may be that, under extreme conditions, it is easier to detect a flock than an individual, but that will very much depend on the observer, the light, the altitude etc. But what are the advantages of monitoring a single animal? If observation is reduced to just one individual in a certain area then the information gathered is poor, since all we know is that a certain species of bird was observed in a place on a certain date, at a certain time, under specific weather conditions, but little more… (And, perhaps, an estimated altitude and direction). However, if we know that the bird was *monitored* on a particular date, at a time in a certain place, then the information becomes much more valuable and gives us vital data on its movements. The reader will have by now realized that we are referring to the scientific ringing of birds, a method which was started in 1899 in Denmark by H. Christian Mortensen, who put metallic, numbered rings on some starlings together with his address. Nowadays there is a European Union for Bird Ringing (EURING), which is the co-ordinating organisation for all

European scientific ringing schemes, as well as some in North Africa.

After trapping a bird in mist nets, spring traps or cage traps, the expert ringer must fill in a form – later sent to their country's ringing office – recording the date, time, location, length (wing, tarsus, beak, primary wing feathers), weight, sex, condition of plumage, and if it has accumulated fat, or not, before being released. If the bird is recaptured these data will be consulted, enabling the time taken to cover a distance and kilometres flown to be noted, and also if there has been any change in the data in the time elapsed between captures. Thus, if the data for one bird are extrapolated to those of the species it is then possible to deduce the movements that species has undertaken. But if this method is to be efficient the more birds that are ringed the better, so that there is a greater chance of recovering one. In Spain between 150,000 and 250,000 birds are ringed each year – a normal figure for a country so ornithologically rich – and about 1,000 Spanish ringed birds are recovered and some 3,000 from abroad.

Besides metallic rings, PVC rings are usually used on large size birds so that they can be read from a distance. The digits, letters or combinations of both are usually in colour against a different background to make them easily legible. Another model is also used with combinations of different colours but no symbols.

Yet another method of tracking bird movements is by dyeing feathers, preferably wing feathers. This is very useful for large birds like raptors, whose dyed feathers show up very well, and the dye is easy to apply using distilled water, although the disadvantage is that the reference can be lost when the feather falls off during moulting.

Finally, there is wing tagging on large birds. This consists of coloured plastic strips attached to the wing between the tertiary and scapular feathers. They are lightweight, easy to install and they do not hinder the birds.

Thanks to modern technology we now have radio tracking, which consists of fitting a tiny radio transmitter to the back of the bird by means of a light harness. They are battery - or solar-powered and emit a radio signal between 140 and 200MHz, which is captured by the antenna of a radio receiver operated by the ornithologist. The difficulty with this method at the moment is that the bird must be about 15km away from the receiver when flying, although this distance can be less, the nearer the bird is to ground level. Of course, sometimes geographical "accidents" can impede reception of the signal. If the bird's

Rings for White Storks: coloured plastic -readable from a distance- and metallic. Ringing campaign in 2004 in the province of Cádiz.

Booted Eagle with radio transmitter attached to its back. The aerial is visible under its tail as it crosses the Strait.

route can be predicted, then it is useful to be situated on its course to facilitate signal reception. It is a good system to have in the Strait because soaring birds are almost bound to pass that way. This method was first used in 1964 with a Bald Eagle (*Haliaeetus leucocephalus*) in order to learn more about its movements and habitats. The transmitter is normally fixed to a bird's back or on the tarsus (storks and cranes). It is now in general use and has been very successful in the control of mammals, such as the lynx (*Lynx pardinus)*, an endangered species in Spain, fish, reptiles etc. This system is especially useful for large birds, but for the time being not for small, lightweight birds because of the weight of the

The Black Stork is one of the trans-Saharan migrants we know most about, thanks to satellite tracking.

transmitter, harness and antenna. It is now used on smaller raptors like the Booted Eagle, as we have seen on occasions in the Strait.

A great revolution, in other words a system that researchers and ornithologists dreamed of, has become reality with satellite tracking. This system also involves a transmitter whose signal, however, is captured by a satellite orbiting at 1000 km. The satellite sends the information about a bird to a reception centre, which in turn passes it on to a data processing centre. The resulting data is then passed on to the ornithologists organizing the study. Finally, the latter may publish the information obtained via the Internet and update it every now and then. With this system it is not necessary for the bird to be situated a few kilometres away; a scientist – and any curious member of the public – can find out the exact position of the bird in real time and follow it carefully, step by step. Radio transmitters

weigh between 15 and 20 grams – the officially recommended weight is no more than 5% of the bird's body weight. When placed on a bird using a harness, the transmitter may last about three years or so. Satellite tracking has revealed bird behaviour, which had never been dreamed of with the previously available technology, as in the case of the Black Storks from Eastern Europe in which two juveniles and the female adult migrated via Israel and the male and another juvenile via Gibraltar. One of the great advantages of the method is that one no longer has to depend on the re-trapping of ringed birds, or, in other words, pure luck. Now the main problem is how long NiCd or NiMH batteries can last, but this is being solved, in large birds, by self-recharging with solar energy. Imperial, Booted and Golden Eagles, Ospreys, Egyptian Vultures and Honey Buzzards, among others, carry these precision instruments. The second problem is the minimum size of this "backpack" – ornithological slang for the pack consisting of transmitter, antenna and harness – which is rather heavy and bulky to be carried by small birds. It is a problem which miniaturization will probably solve as demand for it increases, in a similar way to the changes in the world of PC's. But, for the time being, we shall have to make do with the traditional ringing methods for small birds.

Satellite tracking is a revolutionary system and efficient too, but it is also very expensive and is only being used on birds threatened with extinction. In Spain it has been used in research into Lammergeiers, Imperial Eagles (both wild and reintroduced), reintroduced Ospreys, Egyptian Vultures etc. For birds the size of White Storks, ringing of nestlings is still popular in this country, despite the fact that they are perfectly capable of tolerating the weight of a radio trans-

Manuel Barcell and Cristina Ballesteros ringing White Stork chicks at Casa de las Bombas (Jimena de la Frontera).

mitter. In Holland, on the other hand, where the stork population is scant in comparison, satellite tracking is being used wholeheartedly with these birds. Perhaps one of the most popular tracking projects in Europe was African Odyssey, organized by Czech Radio, and was followed at the end of the 90's by a multitude of ornithologists keen to learn about the African destination of some Black Storks. The development of this tracking method is providing hitherto unknown data about the migratory behaviour of different species of soaring birds, some of which is quite astounding. We shall come to this in more detail when we look at different migrant species in chapter VI.

The reader must be wondering if anything can go wrong with these detection systems that keep us informed, especially via satellite tracking, of the course the birds follow, because no system can be perfect. Indeed, what we do know are the migratory routes, dispersion of juveniles and death rates; some transmitters also feed back information on weather conditions like humidity, altitude, and force and direction of wind. But we still need to know about weight, plumage condition (moults), biometrics, survival rate, (given that most batteries are of limited duration compared to the longevity of, say, soaring birds) and, in general, those data we acquire from *a bird in the hand*, as the saying goes.

Birds in flight: special characteristics of the Strait

So far, when referring to birds passing over the European coastline of the Strait, we have not given any exact positional details, but from now on we will have to be more precise and to do this we are going to divide the European coastline of the Strait into sectors or zones. Accurately

choosing these coastal zones is important to enable us to define passage vectors, sedimentation etc., without having to resort to very precise locations, difficult to find on a map.

In the three most relevant studies carried out so far, in which large groups of observers took part, specific zoning or sectoring – different authors use different terms - has taken place. In the study made by Bernis, the European coastline of the Strait is divided into two sectors, the western (from the Valdevaqueros inlet as far as the Tolmo inlet) and the eastern (from the Tolmo inlet as far as Atunara beach), subdivided into five parts, of which three are in the western and two in the eastern sector (one between Tolmo and Punta Carnero and the other from the Rock of Gibraltar as far as Atunara in La Línea de la Concepción). In the Courses on Vertebrates in the Campo de Gibraltar given by Fernández Cruz, the coastline was divided into four zones; the first between Punta Carnero and Guadalm-

Sub-sections of the Strait according to bird movement :

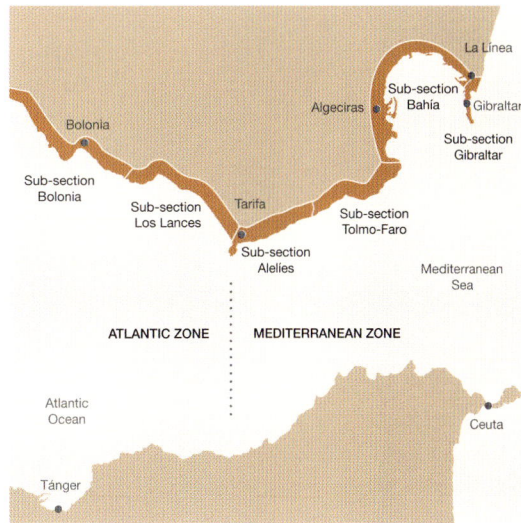

esí; the second between Guadalmesí and where Los Lances beach begins at the east end, next to Tarifa Island; the third between Tarifa Island and Valdevaqueros inlet and the last zone from there to Punta Camarinal next to Bolonia beach. Finally, the Migres Programme divides the coast up into seven sectors, from A, which is Bolonia, to G, which is the Rock of Gibraltar.

We are going to propose two zones: the Atlantic and the Mediterranean. The Atlantic zone coincides with the Atlantic coast of the Natural Park of the Strait, from Cabo Gracia as far as Tarifa Island, in turn split into two sub-sections: in the east, Bolonia, going from Cape Gracia to Valdevaqueros inlet, and, in the west, Los Lances, going from the beach there to Tarifa Island. The Mediterranean zone coincides almost in its entirety with the Mediterranean coast of the Park and goes from the eastern side of Tarifa Island as far as the Rock of Gibraltar. It has four sub-sections: Alelíes, between Tarifa Island

Observatories in the Strait of Gibraltar :

■ The author´s proposals for new observatories

and the mouth of the river Guadalmesí; Tolmo to Faro, between the Guadalmesí river mouth and the lighthouse at Punta Carnero; Bahía (the Bay) and then finally the Rock of Gibraltar sub-section. Why have we given names to the zones and sub-sections? The reason is that we think it is easier for the readers to familiarize themselves with a name that is easy to find on a map, and we are adopting the zoning system suggested by Fernández Cruz because visually it is simpler and more useful in determining the limits of each zone.

In order to make the long hours of waiting more bearable under the Andalusian sun, fortunately somewhat less severe here than in other nearby locations, the Andalusian Regional Government has had several observatories specially built. They are made of stone and cement and probably vandal- and even earthquake-proof. Some of the locations, however, are, in our opinion, somewhat questionable. We consider that one of the most important sites, not only in the area, but also in the whole of Western Europe, is Cerro de Cazalla and there should have been an observatory there. Some of the others are of hardly any use, but the ones that are, are really magnificent observatories. As the observatories were designed exclusively for postnuptial migration most of them are of no use for prenuptial migration except for the ones at Guadalmesí and in the Sierra de Cabrito. So, to meet the needs of ornithologists who come in increasing numbers each year to watch prenuptial migration, at least three more observatories are required: the most important one next to the Punta Carnero lighthouse in Algeciras; another one near Torre del Fraile between the beacon tower and a bunker or even on the bunker itself; and one near the only row of wind turbines in the Park at La Hoya.

The European and Moroccan coastlines next to the Strait of Gibraltar are shaped like an hourglass and one has the impression that the migrant soaring birds coming from the interior are being brought together along the coast and led, like grains of sand, to the funnel of the Strait itself, at which point they inevitably have to decide when to cross over. This applies to both shores. In this way, while one chamber (= continent) of the hourglass is emptying, the other is filling up. As we have already mentioned, this behaviour is due to the flying technique of soaring birds, which, in order to progress further, use thermals and steer clear of vast expanses of water where there are no thermals or these are so slight they have to flap their wings to make progress. But the flying capacity of soaring birds must not be underestimated, as there are birds which in passage reach the Balearic Islands and other islands in the Mediterranean such as Corsica, Sardinia, Malta, Crete and Cyprus, albeit in inferior numbers to those which fly via Gibraltar, the Bosphorus and in the centre via the Strait of Messina between Sicily and Tunisia.

The migrants' approach to the Strait

First of all, let us look at how migrants go about approaching the Strait, because the timing of the crossing will depend on the prevailing weather conditions in the area. The initial position that a bird takes up when approaching the Strait will determine the way it goes about crossing.

From their territories of origin, birds aim for the Strait, trying to fly in a straight line and only altering their flight when faced with storms or heavy rain, or to avoid great heights such as the Pyrenees, which would force them to fly thousands of metres higher. Therefore, birds coming from Europe or from the east and west of the

Iberian Peninsula tend to follow the coastline as they approach the *funnel*. Those from the centre of Spain fly due south towards the Strait to save time and energy. A major physical obstacle for them near Gibraltar is the series of *sierras* in the Natural Park of Los Alcornocales, which, oriented north-south, splits the area into two halves. Migrants try to avoid flying over the top of these *sierras* and prefer to fly round them until they reach the coast.

Headwinds, tailwinds and crosswinds in the Strait

Let us have a closer look at birds' behaviour in accordance with the line they take and the direc-

The effect of winds in the Strait on migrating birds :

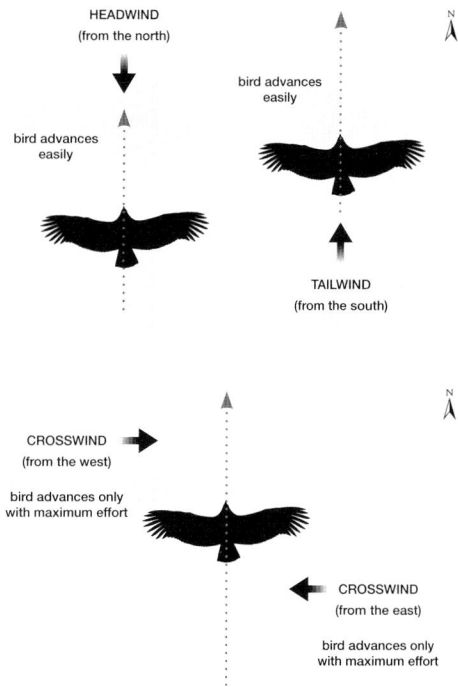

tion and strength of winds in the Strait – be they head- tail- or crosswinds – because these parameters determine the progress the birds make. A headwind is what a bird meets when the wind is blowing in the opposite direction to that in which the bird is flying. In such conditions satisfactory progress is made because, together with their aerodynamic shape, their plumage deflects the wind and reduces friction to a minimum. Birds use winds to rise almost effortlessly and then glide gently downwards. To rise they extend their wings fully, so that the wind pushes them upwards like a kite, and to progress they fold them partially and, depending on the degree of folding, they fly downwards at varying chosen speeds. When the wind is blowing in the same direction as that in which the bird is flying, it is called a tailwind and it transports the bird with hardly any effort, although when this wind is extremely strong it can destabilize the bird. Crosswinds, however, are the most common ones in the Strait, affecting both birds and the relief. Winds from the west and east blow at right angles to the axis formed by the European and African coasts during both pre- and postnuptial periods. Of the three types of wind described it is the crosswind which most destabilizes birds and is the cause of a reduction in the number of migrants crossing and may even bring migration to a standstill.

These winds affect migrants during their two annual passages in different ways. In the area of the Strait the number of days that a north wind is blowing is practically insignificant, but northeasterlies and northwesterlies are more common. On such days, at the end of December and in the first half of January, it just so happens that these north winds coincide with the prenuptial stream of White Storks towards the European coast of the Strait. If numerous flocks of White Storks can cross on days when there are strong crosswinds then quite clearly these northerlies pose no difficulty at all. It would be exceptional for northerly winds to coincide with the migration of Short-toed Eagles and Black Kites but, given that navigation would be into a headwind, there would be little or no deviation from the south – north direction they take from the African coastline. During postnuptial migration it is even more unusual for there to be northerly winds but, if there were, then it would be to the migrants' advantage as they would be *pushed* towards their destinations without using much energy and would keep to their course.

Winds blowing directly from the south (not southeasterlies or southwesterlies) are almost

Drifting of migrants according to force of crosswind :

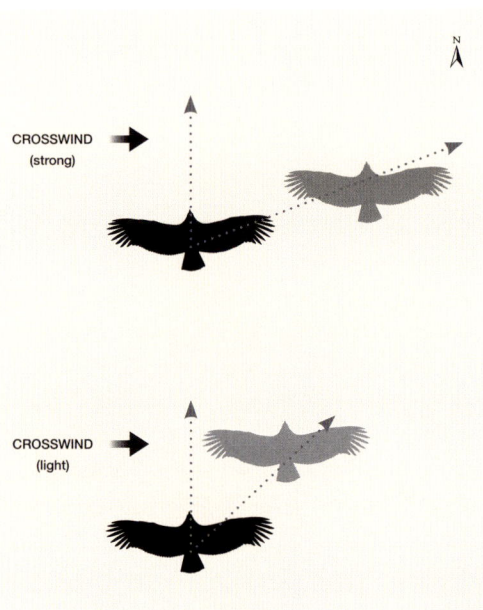

non-existent in winter and uncommon in summer. But, whether they blow in the opposite direction to or in the same as that of the migrants, they help their progress in the same way as the north winds do. In summer a wind from the south makes temperatures rise considerably, but this does not seem to influence migration, at least to any noticeable extent.

Drifting migrants

The main prevailing winds are those blowing from the east and the west, sideways on to the north – south position of the Strait and nearby *sierras*. In addition they bring most of the stormy weather, especially the *levante* or east wind. As we mentioned in Chapter I these winds condition the lives of plants and animals, including man, in the area influenced by the Strait. Thus they must also affect in a big way the migrants which cross it and in whichever direction. Birds tend to cross the Strait in a straight line to make

Drift of migrant over the Strait :

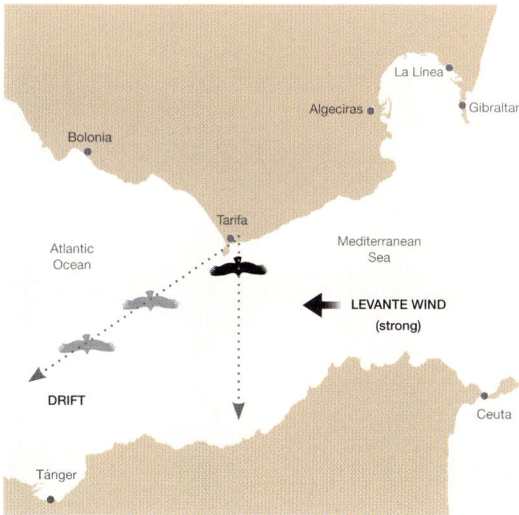

both the distance and time flying over the sea as short as possible, and in order to avoid accidents due to tiredness – in the case of some species like Griffon Vultures to avoid complete exhaustion and drowning in the permanently rough waters of the Strait. The desire of a migrant to keep on course and the deviation caused by crosswinds is called drift. The angle of drift caused by the wind will be greater the stronger the wind is, the more it blows at right angles to the bird's course and the less strength the bird has. This drift can be perfectly appreciated in prenuptial migration from the European side of the Strait when, with a force 6 wind on the Beaufort scale blowing you can see flocks of kites progressing but being dragged by the wind away from the point they are aiming for. These crosswinds are, without any doubt, those that most affect the progress of migrant birds and the cause of a reduction or even total breakdown in the migratory process when they blow with hurricane force.

If we compare the relief of the western coast of the area of the Strait, both on the Spanish and Moroccan sides, we can see that on the African side the coast comes to a sharp point. Therefore, if, during postnuptial migration, a bird is forced to drift westwards because of a strong *levante* wind, it will not find anywhere to land and will end up over the Atlantic where its chances of survival are minimal. On the other hand, if during this same crossing it is dragged in the direction of the Mediterranean by strong *poniente* or east winds, then it will be able to land in northern Morocco. From this it can be concluded that under equal conditions of strong crosswinds, either from the east or west, then birds will be more inhibited by a *levante* than by a *poniente*. This fact must be recorded in migrants' genes and passed down from generation to generation,

Flock of Black Kites flying low and trying to overcome strong gusts of levante near Tarifa.

in the same way that juveniles locate wintering areas without being guided there by adults. When strong crosswinds are blowing the total surface area of a bird may have a very negative effect on its progress, because it performs like a kite and apart from drifting it will also be blown upside down and destabilized. Only the great skill of birds like ospreys and harriers allows them to get out of these situations. Often you can see Short-toed Eagles, Griffon Vultures and some Booted Eagles arriving absolutely worn out, gasping with beaks open and flapping their wings very slowly.

Accumulations, hold-ups, reversal of migration, returns, abandonment and crossing

During the autumn passage, if the wind is from the west migrants will have been subjected to drifting and will approach the Strait from the Mediterranean, or Costa del Sol, side. They will fly along the coast until they reach the Rock of Gibraltar, then go round one side of it and, after crossing the Bay of Gibraltar they will carry on to a place near Tarifa, say Alelíes, where they will wheel and start flying over the sea. If the wind is not too strong for the migrant in question it may even start its crossing from the Rock. However, if it is a stormy wind then the migrants will start accumulations in the Mediterranean section of the Natural Park of the Strait, along the Algeciras coast. The same applies to the *levante* wind but the accumulations will be in the Atlantic section of the Park, along the Tarifa coast. Here they will stay floating with a headwind, making slow progress as they are battered by it, losing and gaining altitude and being pushed back into the European or African hinterland, according to the migration in progress.

For bird watchers this buffeting of migrant birds seems to be cruel torture, but we doubt if they feel it this way, rather it is probably just human interpretation of a fairly irrelevant event for the animal world. On the African side of the Strait migrants gather around Tangier when the *levante* wind is blowing and between Alcázar-Seguer and Ceuta when there is a *poniente wind*. It is these areas that they start from in prenuptial migration.

An exceptional time is when winds blow at more than force 7 on the Beaufort scale – especially *levantes* –and passage of migrants comes to a complete stop, while constant streams of migrants continue to arrive along the European coast of the Strait, forming accumulations which get bigger as the hours or even days pass. On many occasions over the years we have witnessed these hold-ups – or bottlenecks in bird watcher's slang – and it is possible to see a tremendous number of birds over the whole area, many of them, such as Storks and Black Kites, on the ground waiting for the wind to die down and continue on their journey.

A description of these conditions in the Strait, and we were there to witness them, was made by Bernis (Bernis, 1973 *Ardeola*). He reported that on 21st August 1972, after several days of strong *levante* winds, more than 7,000 Black Kites and over 2,000 White Storks (a really high number at a time when the stork population was getting lower by the year) had accumulated in the Atlantic sector and only an estimated 400 raptors had succeeded in crossing that day, most of them Egyptian Vultures. At the time a furious east wind was knocking us about all over the place. Joaquín Araújo, Fernando Parra, other ornithologists and I were staying at the Medina guesthouse in Tarifa. Only seldom did one of us venture to look out from the flat roof to see Kites, Booted Eagles and a few Short-toed Eagles struggling against the wind at low altitude. On 22nd August I left Algeciras at 5.30 a.m. to meet up with my fellow ornithologists at 6. The wind had calmed completely during the night and the winding road was slippery from the film of water covering it due to a thick mist, which I had to drive through. When I arrived a the bar where we used to meet before departing to the different observatories, professor Bernis was really nervous and knocking back coffee as if it were a soft drink. He quickly ran through the names of those allotted to each observatory and said it was going to be a great day, because with no wind, the migrants from the bottleneck plus others arriving from elsewhere would be crossing the Strait. We quickly made our way to the observatories and readied ourselves to live through one of those unforgettable days, wondering aloud what would hit us. Let us quote professor Bernis' words about what we saw that day: "The following day a notable improvement in the weather produced the most impressive mobilization of birds we have ever seen in Gibraltar: a minimum count of 16,295 raptors and about 2,000 storks crossed the Strait."

How do migrating birds behave when they are held up for several days or even weeks on end?

In September 2006, as I check the manuscript and add on things I believe to be relevant or that had simply slipped my mind before, we have been having strong *levante* winds for about three weeks and migrant birds have been accumulating in large numbers, not yet deciding to cross the Strait to Africa. Many of them haven't eaten for some time and you can see Booted and Short-toed Eagles, Egyptian Vultures etc. on the hilltops looking for food. It is also noticeable that the migrant birds are spread all along the coast, as the further away from Tarifa, the less strong the wind is.

I have observed that when the *levante* is really strong in Tarifa, say force 7, then in Algeciras it will be a couple of degrees less, say force 5. This is confirmed in the local weather forecast for shipping. From the human point of view one could say that in Algeciras the wind is not much of a nuisance and you can just about go out for a walk without being blown over!

Migrating birds also appreciate these differences in wind force and when winds blow for long periods, as they are doing now, the birds fly into the wind along the coast towards Algeciras, where the wind intensity is lower and they stay there in much smaller groups than those around Tarifa. You can see birds being buffeted by the wind all along the Mediterranean sector, but as you move away from Tarifa this buffeting gradu-

Reversals of migration in the Strait :

ally lessens and they start gliding along. Friends from Gibraltar, especially Paul Acolina, have told me that nearly every year, if there is a *levante* between 20th and 30th September, hundreds of Booted and Short-toed Eagles gather and float around the north face of the Rock, and it is quite something to see.

One of the first beginner's mistakes made in 1972 by those of us taking part for the first time in the census of the Strait was to count the same flocks of Black Kites heading seawards several times. They would retreat to Los Lances beach and then fly overhead again. These reversals of migration are common in migrant birds like Black Kites, whose reserves of strength are limited. These get ready for the crossing, fighting with all their might against the strong crosswinds, as the observers on the shore watch them disappear into the sea mist. They jot down that a certain flock is crossing the Strait, when what is really happening is that the strength of the wind is making them drift, so much so that the group gives up the attempt and lets the wind carry them back to the coast they have just left. Here they take up positions once more and start all over again. A typical practical example would be a flock of Black Kites leaving the Luz valley and starting to cross with a strong *levante* wind blowing, flying towards Cazalla and from there to the coast where they disappear into the sea mist. Observers note down the number of birds but, later, unable to cope with the force of the wind, these give up and let themselves drift back to Los Lances beach, where they regroup and try again. If this goes on they will be repeatedly counted as migrants, although if several observatories are manned then word can spread and instead of adding to the total they can be subtracted. What is really complicated is to keep

a proper count of individual birds involved in reversals. Sometimes, during postnuptial migration, they are out over the sea and decide to turn back, too tired or afraid to continue, but, as they are at some distance from land, the *levante* pushes them out over the Atlantic where they perish.

The strength of the wind is not always the reason for not crossing the Strait. There are days when a medium force wind is blowing and a flock starts to cross, only to turn round shortly afterwards and fly the same way back: we call this a return. The flock goes back to where it started the passage without drifting to one side as in the case of reversals. These returns are typical, almost exclusively, of White Storks. If only a minority of a flock returns, after attempting a crossing, we call them abandonments. These abandonments usually involve individuals which do not have enough strength to cross, in spite of the protection of the flock, or they are more stressed out than their travelling companions, or are doing this for the first time and because of inexperience, tiredness and stress they abandon their attempt.

The actual crossing is embarked on or at least attempted when the bird itself feels strong enough, irrespective of the weather conditions. Flying twelve or twenty kilometres, depending on the starting point, over a sea like the Strait, is not the same, from the psychological aspect and that of energy consumption, for a Griffon Vulture as it is for a Honey Buzzard. That is why, in the extreme case of these two species, Griffon Vultures are extremely careful about positioning while taking into account the force of the wind. Honey Buzzards, in contrast, can start crossing from the Rock of Gibraltar to Cape Gracia even when strong crosswinds are blowing from the east or the west. What is common in all soaring birds is to find a good position and altitude because these factors are what determine if it is going to be a routine flight or a perilous adventure. Positioning is important in order to make up for the drift caused by crosswinds. Altitude is vital too, since the higher the altitude at which they start the crossing the longer they will take to reach an altitude at which they are forced to flap their wings. These facts are well known to ornithologists who come to the Strait to watch the passage. In the prenuptial they will see the birds arriving at low altitude on the European side and in the postnuptial they will see them leaving at high altitude to compensate for the gradual loss of altitude as they progress across the Strait.

Winds from different directions in the Strait and redirected gusts in the Bay of Gibraltar

Those of us who spend many hours watching prenuptial migration in the Strait have noticed that winds do not always blow in a single direction at the same time and here is a practical example of this.

From Punta del Fraile where there stands a beacon tower, known locally as Torre de los Canutos but Torre del Fraile on the map, we have seen smoke from a ship's chimney near the Bay of Gibraltar blowing west and another sailing past Tangier but blowing eastwards. Therefore we have deduced that in the Strait certain circumstances cause winds to blow from the east and from the west at the same time, as we have already mentioned when referring to the fish meal factory chimney in Chapter I. Sometimes, too, we have seen the smoke from ships in the bay and from the incinerator in Gibraltar blow-

Short-toed Eagle buffeted by the wind. These gusts make the progress of large birds difficult, but don't stop them.

ing west while that from the ships in the Strait was blowing east. What influence do these cross-winds have on migrant birds?

Migrating birds seem to have a practical sense for saving energy when crossing the Strait, both in taking the shortest way over and recognizing the types of prevailing wind at a given time; at least this is the feeling that bird watchers get. Therefore, we suspect – and it is only a purely personal impression, not backed up by experimental evidence – that they must know about wind conditions and how to use them to their advantage. A good example of this can be seen when Griffon Vultures, in prenuptial passage and faced with a westerly force 6 wind, frequently arrive at the Punta Carnero lighthouse, after turning in the Bay of Gibraltar. Westerlies should take them to Gibraltar, but, surprisingly, they arrive near the lighthouse. This is due to redirected gusts of wind prevailing at the time in the Bay of Gibraltar, which push the vultures in the opposite direction to the one they were travelling in until they reached the area. These birds are subjected to drift but, on picking up an air current going in the opposite direction, they take advantage of these sudden tailwinds.

At other times, birds leaving the Moroccan coast under a prevailing westerly are suddenly faced with a strong easterly in the middle of the crossing, which makes them drift, so that they must make a titanic effort to try and reach the opposite shore. Very often they die from hypothermia or drowning after dropping exhausted into the sea.

Resting areas, roosting sites and feeding stations

Migrant birds cross the Strait at the point where the coastlines are nearest to make optimum use of the energy required to flap their wings constantly over this stretch of sea. Whether they decide to cross or not depends on the time they arrive at the coast and the prevailing winds. If they arrive in the late evening it is normal for them to spend the night in trees in the forest and carry on their journey the next day. This is where the Natural Parks of Los Alcornocales and the Strait are so important as resting areas. The cover provided by the vegetation enables them to rest far from potential threats, usually human; this is but one reason why these Natural Parks must be protected areas. Nevertheless, migrants don't only rest in the forests but also on sandy and pebbly beaches, sea cliffs and even on the ground in clearings, on hills etc. On the Rock, as there is hardly any tree cover, the resting migrants are usually passerines in the shrub land.

In 1972 on the way back from a day trip, together with the well-known French ornithologist J.M. Thiollay, from Tarifa to Barbate – just a day before the spectacular crossing of migrants mentioned previously – we were surprised to spot a roosting site of over a hundred Montagu's Harriers just a few metres off the N-340 road between Tahivilla and Facinas.

These resting areas may be circumstantial, such as the beaches where exhausted birds land after crossing the Strait. We have seen Short-toed Eagles and vultures, which, rather than land, simply collapse to the ground, too exhausted to fly another metre.

So, such sites are not chosen deliberately by migrants. What they do choose are roosting sites. These must be well away from urban developments and human populations, sheltered from the wind and in or near woodland. A few years ago, during postnuptial migration, we came upon a roosting site of Short-toed Eagles in a wood where the birds were roosting well apart from the each other, unlike kites and harriers, which huddle together almost wing to wing. We only know of two species using the same roosting sites year after year: White Storks and Black Kites. All other migrants seem to spend the night in areas they do not return to in later years.

Many migrants feed during their journey and, sometimes, use the same feeding stations each year. In the seventies in August, when the passage of Black Kites was in full flow, I used to enjoy driving up to the top of the hill above Bolonia and then roll down silently in neutral gear, watching the Kites through binoculars, feeding on the thousands of grasshoppers there. I could sometimes get within just a few metres of them while they were gorging themselves on these insects. This went on for a few years until the grasshoppers disappeared. In the early eighties, too, White Storks used to gather in summer in the Luz valley to snap up grasshoppers by the thousand near the road leading to the sanctuary of Our Lady of Light, patron of the town of Tarifa. After these feasts both Black Kites and Storks would sleep nearby at the same roosting sites, year after year.

From the early nineties onwards, Black Kites, White Storks and Griffon Vultures have discovered rubbish dumps where there is a readily available and plentiful source of protein. They stop there to feed both in pre- and postnuptial migration. They could be described as the equivalent of petrol stations for humans: they refuel, increase fat reserves and continue their journey. This guaranteed supply must certainly have had an influence in the survival of many migrants, which, without it, would have perished during postnuptial migration after passing the Sahara desert. The same happens on the return journey when migrants have overcome the Sahara and the Atlas mountains and find a table laid and waiting for them! And this is a guarantee of survival.

The two most important rubbish dumps for migrants are in Los Barrios and Medina Sidonia. Both are on migratory routes and have influenced the course taken by many migrants, which fly to them to load up with energy for their migration journey. Later on in this book we shall see how Griffon Vultures have changed their routes to eat at the Basurero Mancomunado del Campo de Gibraltar, next to Los Barrios, and how they have adapted their feeding habits to what they find there.

Not all soaring birds feed during migration; furthermore, the majority do not feed at all, at least if they are fit. Indeed, one of the clues to determine whether a bird is a migrant or not, is that local birds go looking and hunting for food while migrants do not. If, in September, when Short-toed Eagles are passing, we see one hunting, then it can be automatically be counted out as a migrant. On the other hand, you often see Ospreys, White Storks, Black Kites and Griffon Vultures stop to feed. Ospreys fish in the sea and rivers, vultures feed on African carrion and scavenge in European rubbish dumps, White Storks eat locusts in Africa and urban waste in Europe and Black Kites do the same in both continents. Like everything else in life there are exceptions to this rule of behaviour and, in the Strait, these are closely related to wind strength. When stormy weather persists and raptors haven't been able to cross for days, you will see them flying low over the coastal hills on the lookout and hunting for food.

Roosting Black Kites not far from the Strait. Sometimes, several thousand will get together when migration is in full flow.

They think that storks
are men from distant islands,
who turn into birds at a
certain time of the year
to go far away and then
conveniently return
to their country where they
turn back into men until
the following year.

Ali Bey
Travels in Morocco, 1814

The migration of the White Stork

Can there be anyone who hasn't heard of the White Stork? It is perhaps the most popular bird in western culture and the largest one of those that live side by side with man. It must have been their diet, based on small rodents, amphibians, fish, reptiles and invertebrates, that endeared them to man, who would rather let them keep these annoying little animals in check and thus give them shelter instead of exterminating them. Quite clearly it is a bird used to marshlands, but in many places it has evolved and in Europe it has become adapted to the open countryside. In spite of its apparent tameness, when its reproduction period is over, both adults and juveniles shun human presence, as would any other wild animal.

On a farm called Las Herrizas, near the small village of Facinas (Tarifa), there are seven pairs of White Storks nesting on prickly pear cactus (*Opuntia sp*), which the owner, Juan, tends with care so that they will return to nest each year. When he and his wife are standing a mere three metres from the nests, the storks don't give them a second glance – even though they may be incubating the brood. It is as if the couple were simply part of the scenery. When I approached from a distance to photograph them, moving nonchalantly so as not to arouse suspicion, they immediately went on the alert and flew off the nest when I was 6 or 7 metres away.

Distribution

The White Stork (*Ciconia ciconia)* is a polytypical species, meaning that sub-species exist. In our case there are two, the eastern stork *Ciconia ciconia asiatica* which breeds in Central Asia, and the western stork *Ciconia ciconia ciconia* which breeds in Spain and Portugal, northwest France,

A spectacular colony of
nearly one hundred nests
near the city of Cáceres.

Italy, Belgium, Holland, Switzerland and west of the Elbe river in Germany. More to the east it is found in the eastern part of Germany, Austria, Poland, Estonia, Latvia, Lithuania, the Ukraine, Hungary, the Balkans and most of Greece. Except for Denmark, it is not present in Scandinavia or in the Mediterranean islands, Great Britain or Ireland. In Africa it breeds in Morocco, Algeria and Tunisia and more recently in South Africa. Although there is a so-called "mixed zone" corresponding to Northern Germany, Holland and Denmark, whose storks will go either way, in general White Storks from eastern Europe migrate via the Bosphorus and those from western Europe via Gibraltar, although the reverse may sometimes occur.

In Spain they are absent from nearly all the Mediterranean provinces, Cantabria, the Pyrenees and the Balearic and Canary islands. There are some nesting pairs in the east of Galicia and it has recently been introduced into Girona in the Natural Park of Aiguamolls de l'Empordà.

Biological aspects

These magnificent birds are white except for their primary and secondary feathers, which are black. Their most eye-catching characteristic is their long red legs and bills, which get redder during the breeding season. Unlike the adults, juveniles have reddish brown legs and black-tipped bills. The normal clutch is four eggs, one being laid every two days. Both adults incubate the eggs although it seems to be the female's job at night after taking turns during the day. Incubation lasts from 32 – 34 days and there may be a time difference of ten days between the hatching of the first and last chick. If there is insufficient food available the last-born chicks will

probably not survive, but if they all live they help to guarantee the continuity of the species. The adults collect and then regurgitate food in the nest and the chicks fight amongst themselves to get a share. When the heat in the atmosphere becomes intolerable, the adults provide water to their thirsty, almost dehydrated chicks. Young storks don't fly until they are about 60 days old and then frequently return to the nest to sleep. Oriental Storks (*Ciconia boyciana*) take four to five years to reach sexual maturity, but some exceptional cases of reproduction after two years have been reported, although broods did not survive. In Iberia storks frequently reproduce at three years old.

The Iberian population is the largest in Western Europe and this *Ciconiiform* has been enjoying an upswing over the last few decades. Censuses carried out in mainland Spain – there are no storks on the islands – go back to 1948 when a postal survey was done by sending out letters to all local governments in whose towns it was suspected there might be White Storks. After some re-checking, this first census "produced" a population of 26,000 occupied nests. In 1957 another survey conducted by post gave 12,688 occupied nests and, in 1974, 7,340. This way of collecting data is, of course, totally un-

Breeding pairs of White Storks in Spain :

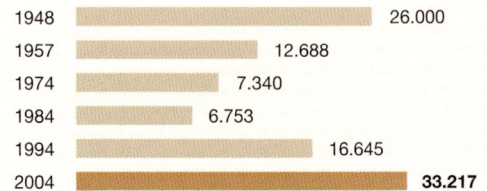

Year	Pairs
1948	26.000
1957	12.688
1974	7.340
1984	6.753
1994	16.645
2004	**33.217**

When migrating, White Storks seek water not only for drinking and preening, but also to keep potential predators at a safe distance.

A spectacular, old-established nest in Ouarzazate, Morocco, situated in the Taourirt Kasbah, a world heritage site, restored by UNESCO.

reliable, because many town councils don't even answer but at least it is a (very) rough guide to the number of nesting birds in Spain. In 1984 and 1994 some proper censuses were conducted and volunteer workers scoured the whole country, as a result of which the official counts were, respectively, 6,753 and 16,645 breeding pairs, showing that numbers had spectacularly more than doubled in a decade. Nationwide censuses have been carried out every ten years since 1974, the last one being in 2004. This produced the impressive figure of 33,217 breeding pairs, doubling the previous total obtained in 1994.

It is estimated that in the eastern part of the European Union including Germany and Poland, which may have about 33,000 breeding pairs – more than any other country – the total population is approximately 550,000 storks (Schulz, 1999). The figures for the west are 29,866 pairs and 112,000 individuals altogether (Schulz, 1999). This latter figure seems to us to be quite accurate because, on passage in the Strait in 1998, the count, reported by Fernández Cruz, was 113,006 individuals at the end of the official census. In actual fact, as he told me, it was about 120,000 by the time he had added those that passed after the end of the census. Furthermore, the total can be added to the considerable number of wintering individuals, especially in Spain. The same author said that in 2004 the estimated number of storks going over the Strait was about 200,000.

This species is not known for its *philopatry* or desire to return to the place it was born, in order to breed. There are some bird species whose juveniles or sub-adults return to the nest but it is rare for White storks to occupy the same territory they were born in, and everything points to them inhabiting unoccupied territories. An

extreme case of this is a stork that was ringed as a chick in Alsace in 1958 and was discovered breeding in Morocco in 1962 (Mayaud, 1965)

Wintering sites of the White Stork

The western European population winters in West Africa below the Sahara desert, mainly in Senegal, Niger, Upper Volta, Ghana, Togo, Ivory Coast and Mali after migrating via the Strait of Gibraltar (S. Cramp, 1977). More recently and possibly due to the ease with which they find food in rubbish dumps or, according to some authors, due to climate change, they winter in Iberia in increasing numbers. The eastern European population migrates via the Middle East (S. Cramp, 1977) and winters in East Africa, from Sudan as far down as South Africa. But, as we are all aware, in the animal world their logic differs from human logic and practical cases belie theories. An exceptional example of this was an Iberian stork whose skeleton was found in 1970 in Kildare (S. Africa), having been ringed as a chick in Tarifa (Cádiz) on 25th May 1969. On the other hand we have western European storks

Flock of White Storks
photographed from a microlight,
near the Santuario valley.

Opposite

Pair of Storks on a nest built on a
prickly pear hedge near Facinas
not far from Tarifa.

that changed their migratory route and joined the flocks of eastern European ones. Moroccan and Algerian storks tend to winter in central Africa (Zaire and Uganda).

In early June 1996 I took part, as a photographer, in a ringing campaign of White Storks, under the auspices of the Andalusian Regional Government, carried out by staff from the zoo in Jerez de la Frontera and supervised by Manuel Barcell. Among the chicks ringed on the Los Derramaderos estate in the La Janda district (where the nests are "guarded" by wild bulls grazing there), there is one, there was one which was later found dead in Tanzania a few months later. The ring was sent to the Spanish consul there and that is how we learned about its journey and fate.

Breeding area of the White Stork :

In a wintering site the size of the stork population depends very much on the seasonal rainfall. More rain means more vegetation cover, which in turn gives rise to an increase in invertebrates and small vertebrates. It is a well-known fact that the White Stork is partial to desert locusts (*Schistocerca gregaria*) and will travel far and wide in the wintering site after swarms of locusts. However, this taste wreaks havoc among their numbers, because in Africa they use heavy doses of pesticides to combat locusts, even in areas where they are not very abundant. The death rate among first year juveniles is between 52% and 74% and among adults 25% to 50% (Bairein & Zink, 1979, Lebreton 1981), according to data based on a study of rings recovered. But the most important factor in the disturbing decrease in population of the White Stork in Africa seems to have been the prolonged drought in Senegal, Mali and Niger from 1968 to 1984 (Dallinga & Schoenmakers, 1989). In addition, poaching and the lack of locusts due to scant rainfall also had a negative effect on the African population, while at the same time in Europe wetlands were drying out, nests were knocked down, more power lines erected etc., making an already delicate situation even worse.

In the Iberian Peninsula an estimated 8,000 individuals winter here, taking advantage of the refuse dumps and, possibly to some extent, climate change.

Migration studies in the Strait

In the introduction we briefly mentioned the studies that have been conducted on migration in the Strait of Gibraltar and we shall now go into those on the migration of the White Stork in more detail.

The first written reference to storks on passage in the Strait was made by Pedro López de Ayala in the XIV century in his *Book on bird hunting*: " I saw in the strait of Morocco, between Tarifa and Ceuta, storks passing at the end of summer, returning to Africa: there were so many that no man could count them, lasting a long time in the sky, so great was the flock." Though not a scientific description, it does have emotional appeal, relating a spectacle of such magnitude that the author is astonished by it.

The general public isn't always aware of events happening around them, especially those that don't have a direct effect on their daily life and more specifically those involving nature. Along with other ornithologists I have come to the conclusion that people don't notice things that are self-evident to us, such as the passage of a large flock of storks over a beach creating a large and very visible shadow. In other words, something quite spectacular. This reminds me of something that happened to me in 1991.

One morning in early August I went into a bar – called Bar España after the owner - on the N-340 road in a hamlet called Casas de Porro, near Tarifa, next to the Valdevaqueros inlet. Señor España was serving a couple sitting at a table, and it seemed to me from their accents that they weren't from the area. I had time to spare, so I wandered over to the door and, looking out towards San Bartolomé in the west, I could see a huge flock of over a thousand White Storks flying towards me quite low above the road. Although I have seen hundreds of large flocks in the area, it was still a breathtaking sight and they were flying so slowly that it almost seemed as if they were showing off. Right away I went back into the bar and said to the couple: "You're not from around these parts, are you? If

you want, come and have a look at something you've probably never seen in your life. You'll be amazed." The three of them came out, the couple and Señor España, to enjoy the scene and, sure enough, they were astounded to be able to see, without binoculars, a great flock of White Storks flying low right towards where we were standing. Señor España turned to me and said: *"Isn't that a pretty sight! In all my time here (*certainly more than 50 years*) it's the first time I've ever seen anything like it."* And it is a fact that every year, over 90% of all the migrating White Storks fly over his bar.

The second account concerning stork migration via the Strait is from Francis Carter's book, published in 1777, entitled *A journey from Gibraltar to Malaga; with a View of that Garrison and its Environs; a Particular Account of the Towns in the Hoya of Malaga; the Ancient and Natural History of Those Cities, of the coast between them, and of the Mountains of Ronda. Illustrated with the Medals of each Municipal Town; and a chart, perspectives, and drawings, taken in the year 1772.* This extensive book title was summarized and in its Spanish translation reads *Journey from Gibraltar to Malaga*. This epigraphy, numismatics and ancient history enthusiast stayed in Gibraltar for a long period from end June 1771 until 23rd September 1772. From his accounts we can select two very pertinent comments, one of which refers to the Barbary Partridge (*Alectoris barbara*) and we quote as follows: *"You can also see (in Gibraltar) red-legged partridge, which were brought from Barbary"*, which infers that their introduction was prior to 1771, when he arrived at the Rock. Another comment is the one he makes regarding the passage of Griffon vultures over the Rock: *"You can easily tell them from storks (which are also passage birds) by their feet tucked*

According to Francis Carter the Barbary Partridge was introduced into Gibraltar in the 18th century.

under their tail, whereas storks fly with them hanging down."

Another Briton, Irby, wrote about migration, although, bearing in mind what we know about it now in the area, his assertions are not very sound, as he said that the autumn passage is much less important than the spring one, when it is exactly the other way round. This must have occurred because on the Rock you can only observe migration, both outward and inward, when the wind blows from the west and when he was staying on the Rock there must have been autumnal *levante* winds. Whenever one of the two winds prevails during one or more seasons, then you will see more or fewer birds from Gibraltar or from any other point in the area of the Strait. This means, for example, that if you stand on Tarifa Island for either of the two migrations and for a whole month there is a

Opposite

The migration of Griffon Vultures across the Strait was mentioned first by Francis Carter in 1777.

The White Stork has a long, strong
beak which it uses to great effect as
a predator and nest builder.

strong west wind blowing, then you will see few birds; the majority will pass over the eastern end of the Mediterranean Zone, including Gibraltar, of course, and even further east if the winds are really strong.

During both passage periods it is very rare to see White Storks over Gibraltar, because it is a long way off their migratory routes, so any sighting there is just accidental. Thus the studies conducted by Evans & Lathbury (1973) are not relevant as far as these migrants go. In the Migres programme in 1999, out of 36,386 storks detected in postnuptial passage from the Bolonia observatory, only 40 individuals were seen in Gibraltar. This gives you an idea of the limited importance of this species' passage via the Rock.

As I mentioned in the introduction, the first study on the migration of the White Stork in the Strait was conducted by Bernis in 1972 as part of a general study on soaring birds. Monitoring, both in that year, in which I took part, and in 1974, lasted from 1st to 31st August and didn't include the second fortnight in July, during which, in some years, a large number of storks migrate. The following details will give an idea of the importance of that fortnight.

In 1976 50,281 White Storks were counted on passage from 16th July to 31st August. Of these, 24,898 crossed between 16th and 31st July, i.e. 49.5% in the second fortnight of July. In 1977 of the 39,162 White Storks counted on passage to Morocco, 22,604 (42.3%) crossed in the second fortnight of July. Thus, the figures given by Bernis – 16,167 in 1972 and 13,450 in 1974 – are not accurate because counting only started on August 1st. However, it could be concluded from these data that between 40% and 50% pass in the first fortnight in August, which

would be a mistake because, for example, in 1978 out of a total of 33,453 storks only 9,084 - 27% - crossed in that period. What is it that holds back or brings forward the arrival of migrants in the Strait? It could be the availability of food or favourable or unfavourable weather conditions.

The above figures show an alarming decrease in numbers of the species but, as we shall see shortly, they are in stark contrast to those obtained by Fernández Cruz, which reflect an amazing and yet unpredictable recovery in the 90's.

In these initial studies the count included not only those birds on passage but also examined the areas of rest - which were quite different from the current ones -, flight strategies, phenology, roosting sites etc. These provided a blueprint for future studies. So we can say that, for the first time, studies in the area quantified the number of migrants crossing the Strait and were useful in determining the logistical strategies necessary to monitor this migration, which varies in relation to those of the other soaring birds.

Numbers of White Storks crossing the Strait on autumn passage :

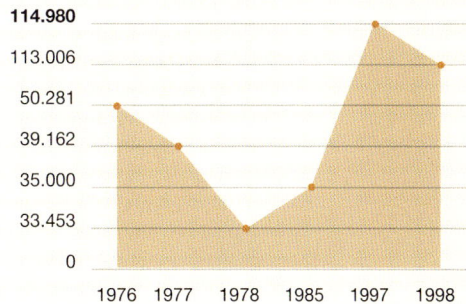

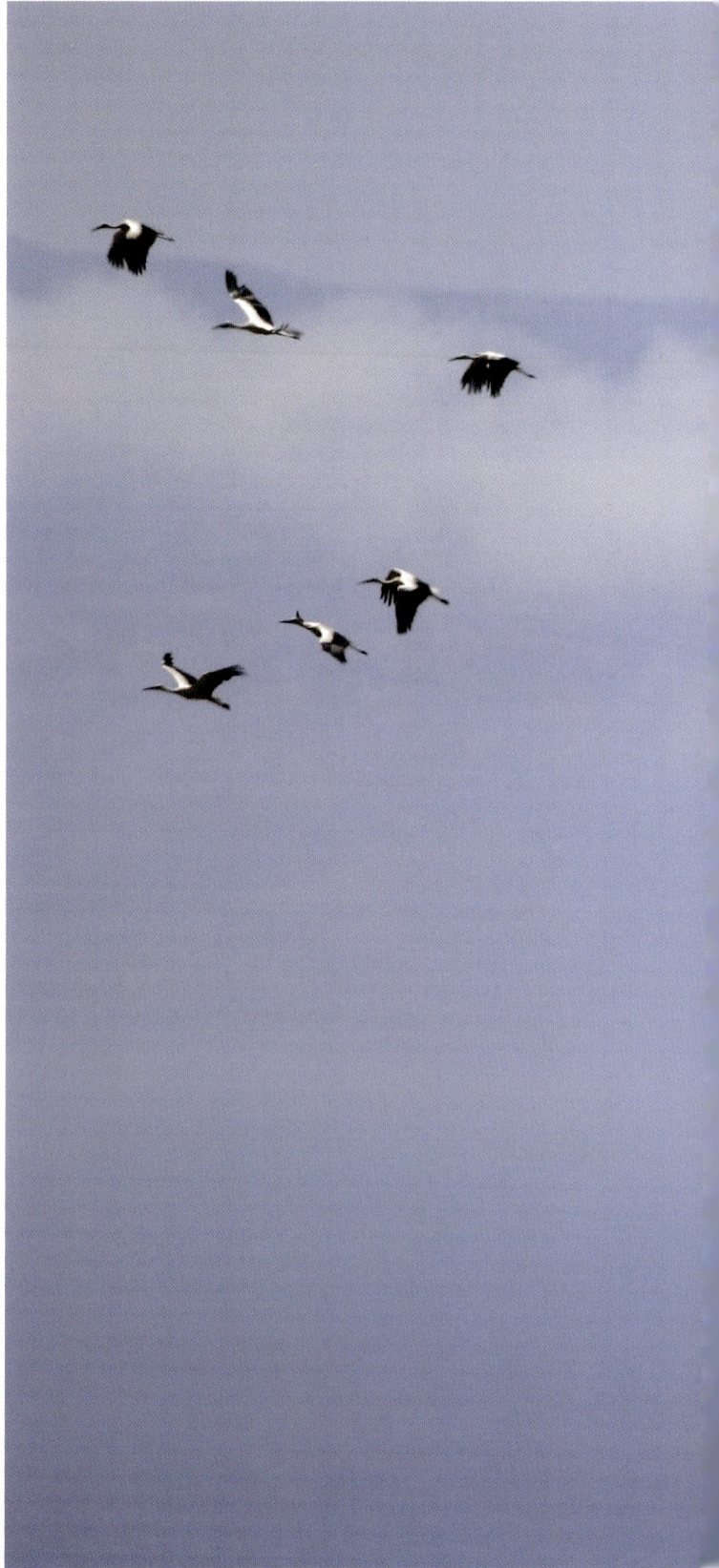

Part of a large flock of juvenile storks flying over the sea, parallel to the European coastline, not having made up their minds to cross.

The next campaign was led by Fernández Cruz in 1985 (*Migration of the White Stork (Ciconia ciconia) through the Strait of Gibraltar in 1985: First results*), subsidized by the Ministry of Public Works, and produced the promising and unexpected number of 35,000 storks on passage. In spite of the efforts, the means available were not quite adequate for a project of this size, as there weren't enough people for the number of observation points needed. There was only one fixed observatory in Facinas, near Tarifa, plus mobile observers in the area who went round trying to detect flocks which couldn't be seen from there. Strategically it is an ideal spot, because it is situated in the north of the *Sierra* Fates, which together with *Sierras* Plata and San Bartolomé, flank the valley of the river Valle leading from La Janda to the sea and it is down this valley where 90% of White Storks fly towards the Strait. Therefore a permanent observatory in Facinas and another in *Sierra* Plata or on the hill above Bolonia would cover the whole valley and you could detect nearly all the White Storks on their way across the Strait to Morocco.

Without a doubt the study which has so far been the most exhaustive, rigorous and longest and which involved the largest number of observers – and I took part every year – was also directed by Fernández Cruz and backed by an agreement between the Complutense University in Madrid and Tarifa Town Council, as part of a Course on the Migration of Vertebrates. It was repeated every year for seven years (1993-1999). Those who took part in the monitoring of postnuptial migration of the White Stork were: Fernández Cruz, as coordinating director – joined in 1993 and 1994 by Dr Valentín Buencuerpo, professor at the same university -,

three assistants, all postgraduate biologists, and between 35 and 40 students, the majority from the Biology Faculty of the university, except for some teachers from schools in the Campo de Gibraltar.

Several publications of this intense field work and painstaking data collection have appeared: a paper read at the III Conference on the Study and Conservation of Flora and Fauna in the Campo de Gibraltar, by Fernández Cruz and Cristina G. Sarasa: *Postnuptial migration of the White Stork (Ciconia ciconia) in the Strait of Gibraltar: Characteristics of the populations in the Iberian Peninsula and Europe*. Recently the Autumn Migration of the White Stork across the Strait of Gibraltar as part of a publication by SEO/BirdLife entitled *The White Stork in Spain, VI International Census 2004*. There were also two reports in bulletins published by the Migres/SEO Programme in 1997 and 1998.

In these studies the official figures for 1997 were 114,980 birds crossing the Strait between 21st July and 31st August. From 1st September until 12th October the Black Stork Ornithological Group (COCN) was in charge of monitoring. During the same period in 1998 113,006 individuals were counted, this time under the auspices of the Migres Programme/SEO, although not with the same diligence and dedication. Besides these very interesting data obtained on birds crossing the Strait which do confirm the unpredictable recovery of the species, the studies also covered migrants' behaviour, accumulations near the Strait, time of passage, segregation of migrants according to age, phenology etc., all of which we can compare with data from the earlier campaigns led by Bernis.

Over the years there have been many amusing experiences such as when a flock was passing and one of the storks regurgitated, the "ball" landing just a couple of yards away from where observers were standing near the water tanks in Tarifa. On one occasion, a stork peeled off a flock to land right next to a group of observers and allowed itself to be picked up. Most of all, however, one remembers the camaraderie present in all these campaigns.

I am not aware of any publication concerning the prenuptial migration of the White Stork in the Strait and, knowing as we do how these birds behave on prenuptial passage, we doubt whether it would be economically viable to conduct one, as it would have to last at least from early December to mid February. Apart from the long time involved it would be very windy, cold, damp and wet and you would need a lot of observers to cover the whole coastline of the Natural Park of the Strait and the mouth of the river Palmones, plus mobile observatories …., in fact a great deal of work, uncertain and inaccurate results and huge expenditure. It would be more economic and pragmatic to monitor arrivals in detail at the height of the migration over a couple of weeks.

For two years from 1996 to 1998 Cristina G. Sarasa and J.R.Garrido, with the aid of Fernández Cruz, made fortnightly counts of White Storks at the Miramundo refuse dump in Medina Sidonia, during which they made an estimate of the maximum and minimum periods of arrival at and departure from the Strait of Gibraltar.

At the time of writing the longest studies conducted into prenuptial migration were those carried out by the Vertebrates Department of the Complutense University in Madrid, led, as in the postnuptial migration studies, by Fernández Cruz, and they were during the first two weeks of May in the 80's. But, although some flocks of

storks were still crossing, they were just incidental. Similar studies have also been conducted in May in the 90's by teachers and students from the Pablo de Olavide University in Seville and more recently by COCN in 2003 −2004, but this is not the right time for the migration of the White Stork.

Prenuptial migration

The reader may well be wondering why this book is almost exclusively about postnuptial migration and hardly touches on prenuptial migration. In addition to the reasons mentioned above there are long periods when not one migrant is crossing and to be able to monitor the whole migratory process of soaring birds, you would have to begin in early December - to be sure of counting the first White Storks – and finish at the end of June, so as not to miss the arrival of the last Griffon vultures on the coast of the Strait. The strong winds in the area cause the migrants to arrive at places which are as far from the Strait as Conil and even Málaga, so that gives you an idea of the enormous task to be undertaken.

Migratory flight

Of all the soaring birds that migrate across the Strait of Gibraltar, the White Stork is the only one to be seen crossing in one or the other direction every month of the year and in smaller or greater numbers depending on the season. It is also significant, in order to become familiar with the migratory strategy of the species, that they prefer flying in compact flocks, and this great coordination among individuals assures them of optimum conditions to be able to progress with considerable saving of energy and a greater chance of a successful crossing. Apart from this migratory sociableness, the White

202
203

A spectacular gathering of migrating
White Storks one evening at the
Comisario lagoon. On occasions
10,000 have been counted there.

Stork is a terrific flyer possessing great power in its wings and is a masterful glider. What we have here is a very powerful bird with enormous stamina, capable of beating its wings for long periods without getting tired, very versatile when gliding and catching thermals in order to rest, go higher and then progress with minimum effort and, finally, a bird capable of forming spectacular flocks where you can just feel the harmony, sense of purpose and − if we may say so − organization. For all these reasons it comes as no surprise that the White Stork is prepared to cross the Strait, come wind, rain, thick fog or whatever. Logically we're referring to adult birds, which in the eyes of expert ornithologists display all these qualities. Postnuptial migration of juveniles will be discussed later on, but in prenuptial migration, at least some sub-adults, returning from Africa for the first time, are able to navigate in flocks where the skill of their elders, already very familiar with the route, gives them confidence to add to that already acquired when they came through the sea corridor separating the two continents.

A curious piece of information about the flying strength of the White Stork is that, when prenuptial migrants arrive on the European side of the Strait, and with the exception of Honey Buzzards, all other species fly in low or very low especially when weather conditions are adverse with strong crosswinds. The way they arrive is directly proportional to the physical condition of the bird: the greater the bird's strength, the higher it will be flying and vice versa. We are, therefore, referring to those migrants which, when it is no longer possible for them to glide, must flap their wings or fall into the sea. We have been observing prenuptial migration for many years but have never seen the White Stork flying below 50 metres when it reaches our shores. This doesn't mean it's impossible, nor even improbable, but it is not frequent. We have seen Honey Buzzards reach the coast flying lower when there is thick fog, but we think this is because of lack of visibility rather than lack of strength.

Prenuptial flight routes

Regardless of the weather prevailing in the Strait, White Storks generally arrive at Alelíes, carry on towards the Valle del Santuario and then on to the Miramundo rubbish dump in Medina Sidonia, where they recharge their batteries. After that they split up into smaller flocks and start dispersing. Some groups arrive via Tolmo and tend to alter their course towards the rubbish dump in Los Barrios. Occasionally I have seen flocks coming in over the Punta Carnero lighthouse and even over the Bay of Gibraltar on days with strong *poniente* winds.

During this prenuptial migration it is quite common to see large groups of White Storks on the ground alongside the river Palmones and in the rice fields where the Janda lagoon used to be, resting and preening before moving on to the dump. These two rubbish dumps nearest the Strait have certainly had an effect on the course taken by the migrants, especially the one in Los Barrios, which didn't use to be on the White Storks' route. Before it was set up, you would occasionally see a group of storks along the river Palmones, but the main body of migrants would come in over Alelíes, fly towards Valle del Santuario, between the *sierras Saladavija* and *Ojén* and into La Janda district. They always fly over plains and rolling countryside, avoiding mountainous land. Up to the end of the 80's you would see large groups of storks on

Opposite

Part of a large flock, about 2,000 strong, circling upwards all together, near the Sierra de Enmedio, Tarifa.

the site of the former lagoon in La Janda, resting after their long journey, preening and eating in the meadows. In the early nineties these groups became less common because they would fly directly from the Strait to the Miramundo dump to refuel. However, now that rice fields have been introduced in the area you can once again see these large concentrations of flocks. We have always wondered why the route chosen by storks on prenuptial passage is different from that on postnuptial. The majority of flocks, when returning to Africa, leave the Miramundo dump, fly across the meadows of La Janda and down the valley of the river Valle, which runs into the sea at Valdevaqueros beach. They very rarely cross the *sierra de Enmedio* and nearly all

Main routes across the Strait on prenuptial passage :

■ route followed by greatest numbers of White Storks
■ routes followed by smaller numbers of White Storks

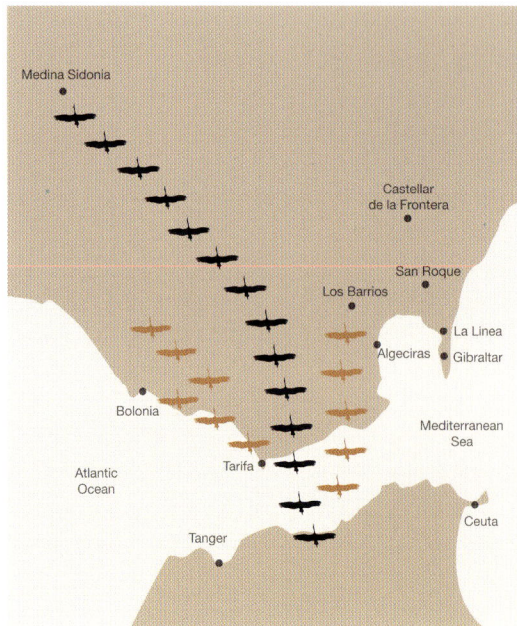

flocks fly parallel to the N-340 road, sometimes flying directly above it, or between it and the edge of the *sierra* or even along the shoreline and over the sea itself. The fact is that they leave the Peninsula over this valley and come in over the Santuario valley. Why does the same stork leave one way and return by another route? In our opinion, although the flocks are not tired on arrival at Alelíes after a good crossing, they want to touch the ground as soon as possible and at the nearest place, so this is as good as any. If once having landed, they want to return by the route they had previously left on (the valley of the river Valle), then they have to fly along the shoreline and even a bit of the way over the sea. Perhaps what they really want is to feel safe flying over land and reject the possibility of flying over a stretch of sea again.

Some flocks returning from Africa come in up the Santuario valley and others, a minority, along the river Palmones; still others fly towards the Miramundo dump in Medina Sidonia or to the one in Los Barrios. But the reader must surely be wondering, as I do, what it is that makes a flock take a particular route. Are these flocks made up of individuals that follow a specific route? Do the dominant individuals decide during the crossing and drag the rest along? We may never know the answer and these questions may stay open to speculation and conjecture, like so many others in the animal world. Nevertheless it may also be that the technical means to answer these questions do exist, but are too costly at present.

Effects of drifting on flocks of Storks

How does drifting affect White Storks on prenuptial migration? In the case of these birds

■ route followed by greatest numbers of White Storks

■ routes followed by smaller numbers of White Storks

drifting does not take them very far from the central area of the Strait (Alelíes). As already mentioned they fly in compact flocks and are birds which can keep flapping their powerful wings – their greatest asset – showing tremendous stamina and resourcefulness, so that the strong crosswinds in the Strait hardly move them from their chosen course. It isn't surprising, then, that groups of White Storks rarely come to the Rock of Gibraltar on prenuptial passage, although there are more than on postnuptial. This drift to the east may exceptionally take them beyond the mouth of the river Guadarranque, and if there is a drift to the west we have only seen a few flocks entering via Bolonia, although some may possibly go as far as Cape

Gracia and one or two even a bit further in very stormy weather.

At mid-day one sunny Saturday in the month of January I was having lunch outside at a restaurant in Bolonia when I saw a medium-sized flock of White Storks flying towards me. I grabbed my binoculars and watched them coming slowly but surely towards the beach. This made everyone look that way and contemplate, open-mouthed, something they had never seen, nor, possibly, imagined.

Prenuptial phenology

A Spanish saying says "*Por San Blas cigüeñas verás*", meaning that White Storks traditionally arrive in Spain in the first week of February. But, as we shall see, this is not quite right, because the saying is from Castilia, in the centre of Spain, where storks arrive later than in the south. The first arrival we have observed and recorded in our field notes is on 15th August 1997 and we quote: " Cazalla (Tarifa) observatory, 11:00 a.m. A small flock of 14 storks flying high and straight in a northerly direction". You could speculate that it was a case of some stragglers turning back, but those of us who have observed White Storks coming and going across the Strait so many times just know when birds are turning back, circling or arriving. These storks were flying very high, in perfect navigating formation and flapping their wings slowly and harmoniously. We are not including them in the migration category, but they certainly arrived from the opposite shore of the Strait in a dispersive movement.

We have a field note recording an arrival of storks on 4th October 1991 but, on a more regular basis, our field notes for October 1994 include the following data: 17th October 1994: Arrival of

Two juveniles squabbling
over food at the
Miramundo dump.

52 White Storks along the river Palmones (7:30 a.m.), 20th October 1994: Arrival of 22 White Storks along the river Palmones (no time given), 22nd October 1994: Arrival of 5 White Storks along the river Palmones (14:30 h).

I could go on listing more arrivals but I'd rather mention a more recent one in 2005, as it is of interest both because of the date and the time of day.

On the 1st of November near the Molinos stream I saw a group of 60 White Storks flying in, about 50 metres above the ground and going first north and then westwards, skirting the hills forming this small depression in the terrain, before finally coming to rest near a row of wind turbines, in order to spend the night. That evening I was accompanied by my wife, Palma, and Enrique Aguirre, a photographer friend of mine, and we were taking photos of the coastline at dusk for this book. On our way back, at 18:30 hrs, with the lights of our off road vehicle on and between two lights on the horizon, we saw a flock of between 80 and 100 storks arriving at that very same spot, the Arroyo stream. We were able to see them because they were highlighted against the dark red sky, for had they been flying lower against a background of land, we wouldn't have seen them.

There are some authors who think that the real migration of White Storks is from mid January until mid February and that any arrivals prior to this are movements of birds wintering mainly in Morocco, but we disagree with this entirely. For whatever reason, storks start arriving at their breeding areas in October in regular numbers, at least in the south and you can see adults on their nests. In November arrivals are more frequent and flocks are bigger, so that by December you can witness the arrival of typi-

cal large, migratory flocks. Arrivals reach their peak in the second half of December and first half of January and you can see similar numbers of storks on the dumps at Miramundo and Los Barrios to those during the summer migration.

Premigratory stopovers: The Wetlands of Doñana

Once the White Storks' chicks are fledged, they leave the nest, returning for a few days only to sleep. Afterwards, these juveniles disappear from their areas of birth and occupy other territories, which may be nearby or hundreds of kilometres away in any geographical direction that takes their fancy, while the adults, against all human logic, stick to their nests and continue to sleep there. On their travels, the juveniles form little groups with others which have also "emancipated" and even with adults that have either failed to raise a brood or did so much earlier in the season. The current trend is to visit the rubbish dump nearest to their new territory, possibly because they see fellow storks there too. In stork breeding areas where there are no dumps you can see them in groups in fields, meadows, pastureland and wet areas.

One of the places to see great gatherings of White Storks is in the Guadalquivir wetlands, situated in the National Park of Doñana (known as the Coto de Doñana) and the Natural Park of Doñana District. It is estimated that about 80% of all White Storks flying towards the Strait of Gibraltar pass through here. In his book *Bird migration in the Strait of Gibraltar* Francisco Bernis, referring to White Storks in the Guadalquivir wetlands in the 1920's and 30's comments the following: "There used to be duck hunters in the summer months who would go shooting storks, which, once plucked, were sold in village marketplaces as cranes." Nowadays the area still attracts a great many storks but, as we shall see later on, where they really get together *en masse* is near the Strait.

There may be several reasons why White Storks are so drawn to these wetlands, the most notable being the almost total absence of humans, a plentiful supply of food (crayfish), the proximity of popular nesting areas like Extremadura, and the nearness to the Strait. The storks from Extremadura that decide to migrate do so by heading south and find these wetlands on their natural route. So logically it proves to be a populated place during prenuptial migration too. Another, no less important, reason is the presence of water. These birds seek out stretches of water for drinking and bathing purposes.

Migratory routes towards the Strait

Although there isn't just one single, visible route towards Gibraltar the most noticeable one is perhaps the one that joins Extremadura with the Guadalquivir wetlands, from there to the Miramundo rubbish dump in Medina Sidonia and the final push to the Strait. Storks approaching Gibraltar from a northeasterly direction are the ones that stop at the dump in Los Barrios to recharge their batteries before crossing.

One of the peculiarities of stork migration is that flocks arrive at the coast from the north west, whatever the prevailing wind strength

Storks at the Campo de Gibraltar refuse dump with the Rock of Gibraltar in the background.

and direction may be at the time. Any arrivals from the north east are with prevailing westerly winds, but not always, and hardly ever with *levante* winds.

Migration resting sites near the Strait

Flocks which leave the Guadalquivir wetlands early may be able to arrive at the Strait in good time and then cross the same day, weather permitting. However, this is not the only option for migrant storks. Nowadays most of them stop off at the Miramundo and Los Barrios dumps to refuel, the gatherings at the former being absolutely spectacular. There is a constant coming and going of groups of birds of all sizes at any time of day you care to look. Sometimes, for

Migratory routes towards the Strait :

■ White Storks coming from Northern Spain
■ White Storks coming from Northern Europe

SPAIN

Mediterranean Sea

Atlantic Ocean

AFRICA

no apparent reason − from the human perspective, anyway − one or two of them take off and, gradually, more and more follow until they have formed a flock of a thousand or more, in a thermal, and rise to head in a south easterly direction towards the Strait of Gibraltar, disappearing into the distance.

On other occasions these flocks move on to the nearby Comisario lagoon. The gathering of White Storks at this lagoon, formerly the Terry lagoon, is without doubt one of the most incredible sights to be witnessed in the world of migration. From a series of photographs we have counted up to 10,000 storks here at the peak of the migration season, although the average number on the ground at one time is between 6,000 and 7,000. About two kilometres from the Miramundo rubbish dump it forms part, together with the Taraje and San Antonio lagoons, of the Nature Reserve of Puerto Real (Cádiz). It extends over 40 hectares and lies on limestone soil fed exclusively by rainwater. If rainfall is scarce it dries up in summer, so there are no fish in it. Vegetation there is made up mainly of shrubs (*Tamarix tamarix*), reeds (*Thypha dominguensis*), sea club-rush (*Scirpus maritimus*), giant reeds (*Arundo donax*) and common reeds (*Phragmites australis*). The surrounding area is a game reserve with plenty of hares, rabbits and partridge. Although it's only a few metres away from the A-5105 road, leading from Puerto Real to the A-381 Algeciras-Jerez de la Frontera motorway, it can't be seen from the road thanks to a plantation of eucalyptus trees and you can only watch from outside a wire fence surrounding it, a guarantee of peace and quiet for the birds which rest or breed there. The breeding birds are of course closely linked to this ecosystem and include the Little Bittern (*Ixobrychus minutes*), Purple Gall-

inule (*Porphyrio porphyrio*), Coot (*Fulica atra*), Little Grebe (*Tachybraptus ruficollis*), Marsh Harrier (*Circus aeruginosus*), many species of duck (*Anas sp.*), Black-winged Stilt (*Himantopus himantopus*), Lapwing (*Vanellus vanellus*) etc. As a roosting site it is used in summer by many species such as the Cattle Egret (*Bubulcus ibis*), Jackdaw (*Corvus monedula*), White Stork (*Ciconia ciconia*), Black Kite (*Milvus migrans*) when migrating, and many small bird species. Other birds that stop to rest and regain energy there

Grey Heron at the Comisario lagoon, holding a chicken's foot in its beak, after possibly snatching it from a White Stork.

Black Stork trying in vain to land
among a flock of White Storks.

are the Greater Flamingo (*Phenicopterus ruber*), Black-tailed Godwit (*Limosa limosa*), young Squacco Heron (*Ardeola ralloides*), Little Egret (*Egretta garzeta*), Grey Heron (*Ardea cinerea*) etc. Among the accidental visitors is the Pelican (*Pelecanus onocrotalus*), which spent several days on the lagoon next to White Storks and flamingos in August 1998 and, during that same summer, a Lanner (*Falco biarmicus*), which was hunting the pigeons which nest in a hut on a hill inside the reserve.

There is a continual toing and froing of White Storks between the lagoon and the dump throughout most of the day, but the peak times are in the morning and evening. After dawn, when the sun has warmed up the ground and thermals begin to form, at about eight o' clock in the morning a few individuals start to fly and are gradually joined by hundreds of other storks until a flock of between 500 and a thousand or so head off in two directions: to the dump at Miramundo or towards Gibraltar. Those that go to the dump land on the mass of plastic bags after a few minutes, eager to scavenge for food. This is an ideal place to identify individual storks by reading the rings from a distance, enabling one to know the date and place of ringing, migratory route, journey time etc. Sooner or later, depending on the amount of food they find, they will return to the lagoon. We have spent several days in a hide and watched flocks arriving at the lagoon throughout the day; we assume that some come from the dump and others stop off on their migratory route. On the shores of the lagoon you can see a lot of dead storks, a fact that won't surprise anyone who has seen the kind of toxic, indigestible things they eat at the dump. It is common to see them carrying things like bits of plastic, material, chicken feet, string etc., which they wash in the lagoon and then swallow – or at least try to, because sometimes the size makes it impossible. Other birds like the Grey Heron will try to snatch food from them or they find it floating in the water, so it is not uncommon to see a heron with a chicken foot in its bill.

After nearly all the juveniles have passed, migration begins to decrease and the flocks heading for the Strait are smaller and less frequent, although at the lagoon we have hardly noticed any difference in the numbers roosting there. It makes one think that the adults, already well experienced at crossing and therefore under less stress, take migration in their stride and spend a few days eating, restoring energy and getting their plumage in shape ready for the crossing. For anyone not familiar with the mysteries of migration the sight of 7,000 storks at the lagoon may make them think they will cross the Strait the very next day; but it will be pointless to wait as probably only a couple of smallish flocks will leave.

In the 1980's and 90's storks would rest at the site of the former La Janda lagoon, their favourite spot being an estate called Los Derramaderos where wild bulls are bred and there is a colony of nests. The storks felt protected by these wild animals and would feed, preen and sleep on the ground. At that time there was a plentiful supply of cicadas and other invertebrates providing energy reserves for the long journey ahead. There was also a little stream for the animals to drink from, in which the storks would bathe and clean their feathers.

Another resting site for White Storks in the 70's and 80's was the Santuario valley, near the coast at Tarifa. Surrounded by the Enmedio, Fates, Saladaviciosa, Cabrito and Ojén *Sierras*, this valley is out of range of the bustling traffic on what is now the N-340 and is crossed by a narrow road where it's impossible to drive fast. At that time there were tremendous numbers of bushcrickets (*Ephippiger ephippiger*) which inundated the road, fields and crops. It was ideal food for the storks, which possibly came in search of it, arriving in medium-sized flocks. With the introduction of pesticides and insecticides these crickets disappeared, so that by the end of the 80's the White Stork had stopped visiting this valley.

The rubbish dump at Los Barrios, nearer the Strait, is on the White Storks' eastern access route to it and is different from the one at Miramundo in some respects. It is also isolated from traffic and built-up areas, as is logical, and quite near cork oak forests where there are no watercourses of note – just a little stream, near the river Guadarranque. So due to the lack of watercourses the White Storks use the nearby cork oak trees as a roosting site. The importance of this dump lies in the fact that migrating White Storks from central Europe stop here, but we shall come to this shortly. Another feature of this dump is the number of Black Kites that visit, during both passage periods, to feed and rest here before crossing that feared stretch of sea, which for this less versatile species must be a major obstacle. It is also vitally important for Griffon Vultures on postnuptial passage to replenish their reserves, used up on the first part of their journey, leading them across the Strait and beyond the Sahara.

Postnuptial phenology and crossing times

Although it may vary from year to year, in general White Stork migration begins in the second half of July. Only on few occasions have we seen it start before, although in 2004, we saw a group of four White Storks turn back after attempting to cross on 2nd July, possibly because of the thick fog which prevented them from seeing the African coast. So it wasn't migration as such. Migration starts when you see a flock of 12 to 20 individuals, the next day flocks of 100 or more and on the third and fourth days there will be flocks made up of hundreds of birds. This is true for most of the species that cross via the Strait. It can happen that, after several days of passage, there is an interruption and a few days with very little movement, but a visit to the Miramundo rubbish dump will reveal several thousand birds scavenging for food. Then suddenly, after this

break, flocks of birds start arriving in numbers which surprise even the most veteran observers. The peak period for storks crossing via the Strait of Gibraltar is between 10th and 20th August, after which there is a notable fall in both numbers and size of flocks. From the first week of September on, the size of flocks falls and it is unusual to see more than 100 together. As the month passes this figure decreases and flocks rarely number more than 50 individuals.

Crossing times are strictly between 10 a.m. and 6 p.m., regardless of the strength and direction of the prevailing wind. It is the only soaring bird with such a fixed timetable, so that it is futile to watch out for storks either before or

Storks and cattle at the former La Janda lagoon on a day of strong levante winds.

after those times, unlike other migrants such as Honey Buzzards or Black Kites.

Between 10 and 10.30 in the morning a few of the storks of different groups that are strolling on the hillsides, in the dumps, pools, lagoons etc. start to lift off the ground and soon afterwards form a flock, circling and gaining height to set off for the Strait, whether they are already quite near or as far away as the marshlands of the Guadalquivir. Depending on how far they are from the Strait, these flocks will arrive between 10.00 a.m. and 18.00 p.m., the ones arriving first and last being smaller in size.

If a flock has roosted near the coast then it will arrive at the Strait at about 10.30 whereas one which has left Doñana at the same time will appear at about 13.00. Although flocks don't always start off at that time, at the Comisario lagoon I have observed that at different times of the day small or large flocks will form and head for the Strait in what seems, to the human eye, to be something completely spontaneous, because a few storks take off, circle over the lagoon and are gradually imitated by others so that a good-sized flock forms and gradually gains height, but in no apparent hurry. At a certain time one of them will head off for the Strait, followed by the rest of the flock. However, there are occasions when what seems to be a flock forming in the air ceases to be one, as little by little the birds start landing again and begin to preen themselves or just rest.

Approach strategies and strait crossing by flocks of juveniles

Without hesitation I confess myself to be a real admirer of the postnuptial migration of juvenile White Storks and the prenuptial migration of Griffon Vultures. Just as with any human activity, there are always personal preferences when it comes to observing this or that migrant bird and this is one of them. I remember an ornithologist saying to me, somewhat disdainfully, that she hated the migration of the White Stork and what she really admired was that of the Black Stork. Possibly we should put this comment down to her youth, because any type of migration is worthy of admiration, fascination and the joy of being able to witness an event of this nature in the animal world in these times of destruction and insensitivity.

Going against all human logic the first storks to migrate are the year's juveniles. As we pointed out at the beginning of this chapter they differ from adults in that the tip of their bill is black and their legs are reddish brown. In flight these details are not always easy to detect unless the conditions of altitude, light and visibility are ideal; therefore, to know the age composition of flocks the most reliable method is to observe gatherings quietly and from a safe distance, on the ground – at dumps, waterlogged ground, rice fields etc. During the Courses on Vertebrates in the Campo de Gibraltar, which we have referred to several times in this and other chapters, observers would devote part of the day estimating the percentage of juveniles and adults at the rubbish dumps in Miramundo and Los Barrios to find out in what proportions they crossed and when. Let's remember one simple thing: these birds, for the first time in their short existence, are facing the challenge of crossing the Strait under conditions that vary from what you could call a simple pleasure trip to the most terrifying storm a human sailor could imagine, with crosswinds blowing at over 100 km/h. So if we add the lack of "team spirit" (to use a human term easily understood by all) within the flock

Opposite

The White Stork has no qualms about nesting in the centre of large towns like Jerez de la Frontera.

to the individual's inexperience, we can imagine the state of excitement of a young White Stork approaching the Strait with foreboding.

These flocks of young "beginners" are easily spotted because they fly across in the second half of July, before the adults have reached the Strait, and, generally, at a lower altitude than these flocks of "experts". Like the majority of White Storks, these flocks start the crossing at Los Lances, flying at medium altitude between the shoreline and the *Sierra* de Enmedio towards the city of Tarifa. Sometimes a few individuals peel off the group and start circling in a thermal, but hardly gaining height. The rest of the flock may then stop and do the same but at once they regroup and head for Tarifa or a bit further east where the Tarifa Maritime Shipping Antenna and the water deposits stand. After flying over some hilly ground the terrain drops abruptly down to the sea and an already nervous flock is suddenly faced with the shore and an extensive mass of water underneath it. The reactions of these flocks of amateurs vary greatly and depend on the degree of nervousness within the group, the strength of the prevailing wind, the visibility and the altitude at which they reach the coast. Depending on all of these and, probably, other factors of which we observers are unaware, ornithologically speaking the predictable or the unpredictable can happen. Many a time, when they look down from this medium altitude, the expanse of open sea and a slight crosswind are enough to make those at the front slow down and the ones following have to manoeuvre to avoid crashing into them. Then the flock becomes disorganized and storks fly and circle in all directions trying not to collide with each other. So they retreat in disarray, flying back in the opposite direction, spread out on a wide front in their rush to get back. Flapping

Flock of young storks, flying too low to be able to cross the Strait successfully.

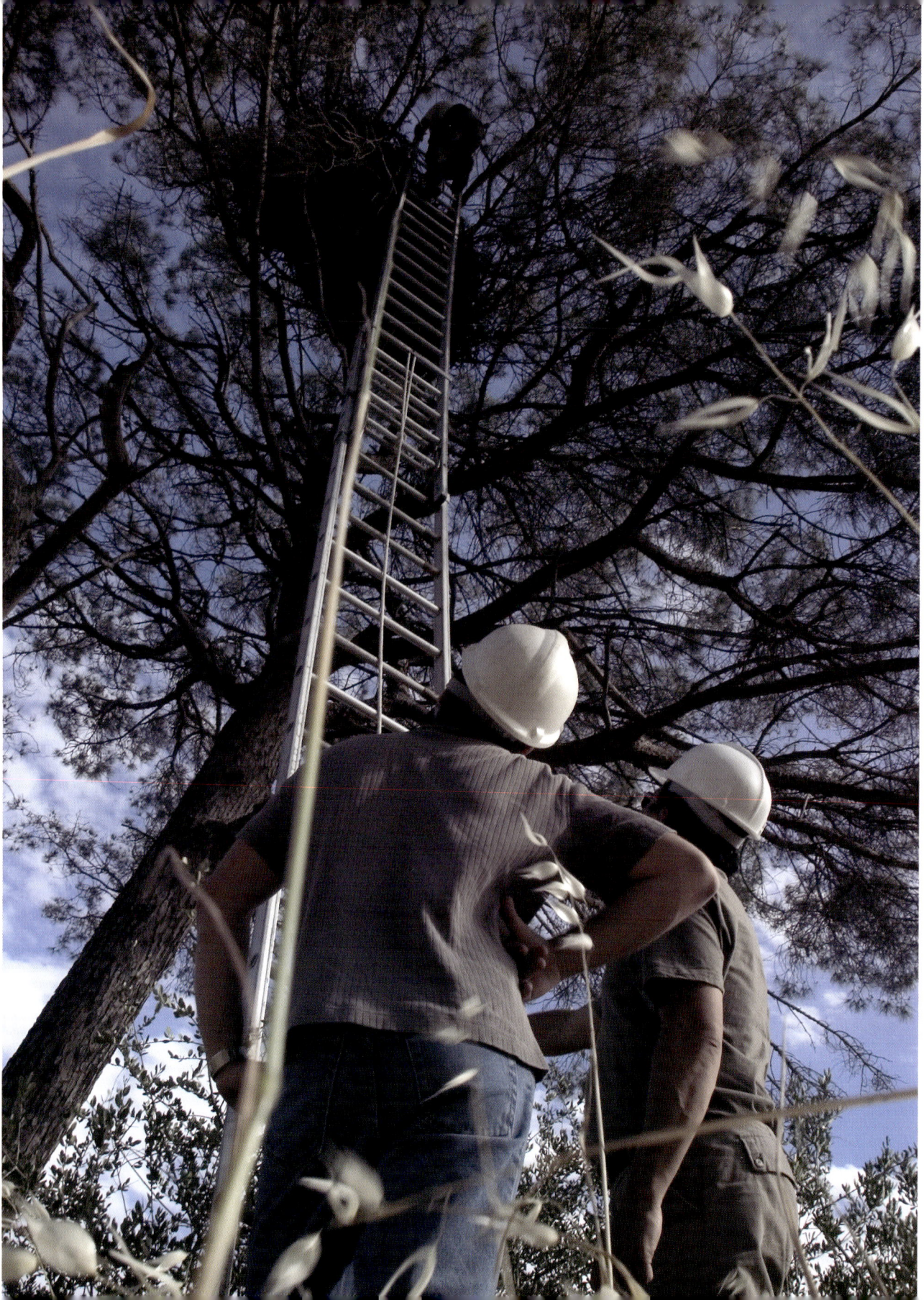

their wings they fly back low towards the Luz valley between the lower eastern slopes of the *Sierras* Fates and Enmedio. Here they regroup and, catching thermals they start going upwards until, at an inadequate altitude, they decide to start trying to cross again, flapping their wings as if enraged and heading for the same point at which they failed before. This manoeuvre may last a whole morning until, physically and psychologically exhausted, they go back the way they came to the area round La Janda and even to Miramundo.

When very strong winds are blowing there may be flocks that manage to advance a good way over the sea, at a suitable altitude for crossing, yet after a few minutes we can see them gradually returning one by one until we have counted the same number that started the crossing. These flocks of hesitant juveniles are one of the great migratory sights of the Strait and show us just how hard it must be for a bird to cross this stretch of sea, even in good weather, when it lacks sufficient experience. As these first flocks crossing via Gibraltar are composed almost exclusively of juveniles there are no adults to use their experience and take the initiative. All the members feel a kind of internal urge that drives them to fly in a certain direction and it is so strong that, after a moment of collective panic, they try again under the same conditions and at the same place.

But it doesn't always happen like that. The fact is that, largely, a successful crossing depends on the altitude at which it starts. We have observed flocks of juveniles flying at medium altitude, turn round when they reach the shore, go back to the *Sierra* Enmedio, regroup on the eastern slope and, taking advantage of slope currents or thermals, soar up. Then they restart at a good altitude, flapping their wings with gusto, and complete the passage to Morocco without further ado.

Weather conditions are definitely a determining factor for flocks of inexperienced juveniles, especially strong crosswinds, fog or heat haze. When the *levante* is blowing force 7 on the Beaufort scale (55km/h) or more, then it is difficult for juveniles to decide to cross and they may not even try but stay squatting on the ground facing the wind. During the first few days of postnuptial migration of Black Kites and White Storks you can see groups of them in the fields waiting for the weather to improve. We have also observed how these flocks shy away from flying over the sea on foggy or very misty days with no wind – a rare event- and we interpret this as a kind of fear of flying "blind" with no African coast in sight. Postnuptial migration coincides exactly with the time of year when there is most fog or heat haze, from mid July to September. This is due to the increase in temperature of the surface water of the sea, heated up by the sun over a long period, compared to the temperature of the nearest layer of air. As a result there is an increase in evaporation and condensation, causing thick, persistent fog to form. This sometimes lasts well into the afternoon.

As they are gregarious by nature, separate flocks tend to form one large one. This kind of behaviour is easily seen when, if a flock has already reached Los Lances, another one appears over the *Sierra* Enmedio and will follow in the first one's wake until they join up, changing its course if necessary. Sometimes flocks split up into smaller ones because, while flying overland some individuals decide to soar while others carry on their course and so a new flock forms. If the group is a large one and soars for some time

Opposite

Ringing of White Stork chicks in a pine tree, near Alcalá de los Gazules during the 2005 campaign.

Compact flock of White Storks near the Strait on a day of strong levantes when coordination between them is vital to maintain cohesion.

A flock, well on its way across the Strait, passing an oil tanker on prenuptial migration.

the main flock may get a considerable distance away. Normally the flock which has broken away will follow the other until it catches up and re-joins the original flock. This is quite typical in juvenile migration, due to their hesitant and in-decisive nature.

Behaviour of adult flocks when crossing the Strait

As the days go by and the migration of White Storks via Gibraltar continues, the type of behaviour outlined above becomes less common. The presence of veterans is now plain to see in the strategy adopted in approaching the Strait. If insufficient height was reached at the first attempt, the storks regroup more quickly and have another go at the right altitude, crossing the stretch of sea at the second attempt with no trouble at all. Research carried out by those studying postnuptial migration at the rubbish dumps in the area show that the proportion of adult storks is increasing, so much so that there is now only a token presence of juveniles. Gradually the flocks coming from Miramundo fly down to the bottom end of the Santuario valley (also called Valle de la Luz) towards the Los Lances beach, at the right altitude and on a fixed course, rarely spiralling upwards for long; it is as if the flocks' course had been chosen by those with experience and this was known and accepted by the inexperienced storks (as in similar situations in human life). Once the adults take over the leadership of flocks, these behave differently. They no longer cross near the town of Tarifa or thereabouts, as you would expect, instead they fly over less terrain and start the sea crossing from the Valdevaqueros or Los Lances beach. What is unusual for soaring birds is that some flocks even come over the *Sierra* Enmedio, which is 400m high and fly at high altitude towards Alelíes, beating their wings gently or just floating along on a steadfast course.

One curious fact about this migration is that until the end of the nineties no flocks of veteran birds had been detected starting their sea journey via the Bolonia inlet. In any case nobody was aware that they did. It just goes to show that in postnuptial migration observers must keep their eyes peeled because birds flying over the area are expert navigators and can fly at altitudes that are

not easy to spot, even by the most seasoned observers. The following anecdote is a good example of what I mean.

I was on the beach at Los Lances one day, enjoying the fresh sea breeze blowing gently from the west. The sky was a deep blue with no clouds to refer to. It must have been in 1996 during the Course on Vertebrates. I would normally have been with teachers and students but that day I decided to spend on the beach, not that I'm a real fan of the beach, mind you. At about 11 o' clock in the morning, although I can't be too sure of that now, I saw a gaudy, yellow para glider, gliding very high along the coastline advertising some alcoholic drink or other. Anyway, I stood and watched it because it stuck out like a sore thumb in the deep blue sky, but I couldn't be bothered to get my camera out to take a few shots of it. After watching it through my binoculars for a while, I stopped and watched some other beach "attraction". A short time later I noticed the hang paraglider was right above my head so, once again ignoring my camera, I got my binoculars out. The impressive sight was a real contrast in colour, the deep yellow shining like a jewel against the dark blue, almost surreal, sky. Just at that moment something came into the field of vision of my well-worn binoculars. There, above the hang glider was a huge flock of White Storks standing out even more against the blue sky. I held my breath while the flock glided over the paraglider, hardly any of the storks beating their wings as they passed it silently. I didn't take a photo but that image is perhaps the one that is most deeply imprinted on my memory. On that day not one observatory recorded the passage of any flocks towards Morocco and it is therefore an example of how difficult it can be to detect the presence of these birds under certain light conditions – so you can never be absolutely sure of monitoring everything that flies over the Strait!

These flocks of expert migrants start the passage some distance away from the Strait and don't need to do so from near the coast. Many of them come from the Guadalquivir wetlands, which is 120 km from the Strait as the crow/stork flies, and, if they leave at 8:00 a.m., depending on wind strength and direction and flying at 50-60 km/h, can reach the Strait in two and a half hours or less. So, at about 11:00 a.m. at the latest they are crossing the Strait of Gibraltar. If they leave from Miramundo it is 58 km to the Strait and they can be over the coastline in an hour. Even the juveniles, whose flight is more hesitant and erratic, can reach the Strait in just over an hour from Miramundo and perhaps in three and a half hours from Doñana.

Migrating routes over the Strait, birds classed by flocks

I have already commented that the great majority of migrating White Storks arrive at the European coastline in the Atlantic sector. The study led by Fernández Cruz reveals that in 1995 95% of these birds migrated via the Atlantic sector, in 1997 the figure was 98%, in 1998 99% and in 2004 it was 97%. This gives us an idea of the importance of this area as an exclusive one for the passage of the White Stork. The remainder come from the rubbish dump at Los Barrios and normally start their Strait crossing at Tolmo - Faro at the mouth of the river Guadalmesí, but so few as to be just a token amount, even though there may be more in prenuptial passage – but that is just a visual estimate, not that any study has been done to confirm this.

According to Fernández Cruz (1995), 95% of all storks crossing the Strait did so between Alelíes and Bolonia. 58% crossed from Alelíes, 25% crossed from Los Lances and 12% over Bolonia. The rest – 5% - were those storks coming from the east and crossed via Tolmo-Faro, after leaving the Los Barrios dump. Another study by the same author puts at 97% the number crossing the Strait between Alelíes and Bolonia. 59% of them crossed via Alelíes, 16% over Los Lances and 22% over Bolonia. The other 3%, coming from the east via the Los Barrios dump, went over Tolmo-Faro. Comparing both sets of figures the only variations are those of Los Lances and Bolonia, but if we add them together there is little change, 37% in 1995 and 38% in 2004.

At the beginning of migration in both directions flock size is small but it starts to increase quickly and after a few days there are hundreds of storks in a flock. In postnuptial migration the largest flock detected so far was in 1998, with about 6,000 birds. It was an impressive sight, as you can imagine, forming an enormous column that stretched from the area of La Peña as far as the isthmus of Tarifa. Although we witnessed it, our photographs were unable to reflect the magnitude of this incredible spectacle because of the light, the flying altitude and the wide front covered by the flock. In 2005 I saw something similar but this time the flocks were spread out along the coast, there were storks all over the place and at one time there must have been some 12,000 or so in the air.

The migration of the White Stork via the Strait of Gibraltar is a sight that is second to none from an ornithologist's point of view and even for the casual tourist. To see a flock of 800 birds spiralling upwards in orderly fashion, gaining height, stretching the shape of the flock, splitting up and then regrouping to catch another thermal to start the process all over again, is a unique and very moving experience. In the Tarifa area you can sometimes observe flocks made up of more than two thousand birds, although the average flock size in, say, 1998 was 264 (Fernández Cruz, 1998).

Between 1974 and 1976 Pineau and Giraud–Audine conducted a study of the postnuptial migration of raptors around Tangier. However, the first data on the passage and arrival of flocks of White Storks in Morocco was collected in a joint effort by students on the Course on Vertebrates in the Campo de Gibraltar – in the, at the time, not yet created Natural Park of the Strait – and members of the Black Stork Ornithological Group (COCN) on the Moroccan coast. This took place for the first time in 1999, but studies have been intermittent. The observers on the Moroccan side were informed by radiotelephone of the passage of flocks from the European side of the Strait by giving the altitude, position and set course. On days with good visibility flocks could be observed from both sides simultaneously; the maximum recorded speed was 96 km/h and the average slightly over 6 km/h. The greatest drift recorded was 22 km from departure to arrival point on very windy days. Recently members of COCN have continued these studies on the Moroccan side, although they do not cover the whole migratory period.

From studies carried out in Morocco exclusively by COCN members we can see many similarities to the arrival of flocks on prenuptial passage in the Natural Park of the Strait. When they get to the Moroccan coast they seek out valleys, just as they do on the opposite coast, so as not to have to fly at any great height.

Postnuptial crossing of White Storks, divided into sectors. :

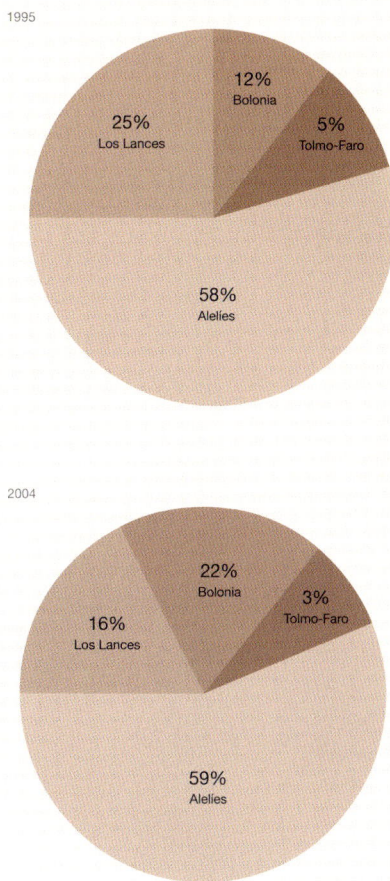

1995

- 12% Bolonia
- 5% Tolmo-Faro
- 25% Los Lances
- 58% Alelies

2004

- 22% Bolonia
- 3% Tolmo-Faro
- 16% Los Lances
- 59% Alelies

Part of a flock of White Storks
arriving at the European side of
the Strait in December 2005.

Stork at the Comisario lagoon wearing
a ring, readable from a distance, on its
right tarsus, and another metallic one
on its left.

Here too there are differences in behaviour of migrants in both migrations. We are referring to their flying altitude when they arrive at the coasts. In the first half of August, according to COCN records, many flocks arrive at the Moroccan coast only a few metres above ground level. (Migrations from the other shore: The migratory behaviour of birds, 2004). However, in our observations on the Spanish coast we have never seen flocks arrive flying below 50 metres high. The explanation may lie in that the prenuptial migrants are veteran birds while the majority of those on the postnuptial passage under study could be juveniles tackling the Strait crossing for the first time: at the time this study was underway (the first half of August) there are always a lot of juveniles still crossing. A further difference in the two migrations is that some members of flocks arrive at the Moroccan coast up to four minutes later than the main body, having left at the same time, whereas those flocks arriving at the Natural Park of the Strait do so in a compact group with hardly any stragglers. The explanation must be the same as before: the storks' experience. COCN members have also reported that the arrival of storks in Morocco covers a band of up to 4 km of coast because of drifting, but, in contrast, in prenuptial migration the flocks are compact and their extension depends on their numbers. It would be interesting to observe arrivals on the Moroccan side of flocks composed entirely of adults from mid August onwards.

The Storks of West Central Europe; special features in migration

It's not only Iberian storks that fly over the Strait. Those we previously referred to as west-ern European storks do so too and the reader will remember they come from France, Belgium, Holland, Switzerland and Germany west of the river Elbe, and also in smaller numbers from what we call the mixed zone.

The first difference between the migration of these birds and the Iberian ones is that they arrive in the Campo de Gibraltar later and settle at the Los Barrios rubbish dump. They come down over the Mediterranean provinces of the east side of Spain and when they arrive in the area they find this dump, possibly – and it's just a personal opinion – drawn to it by the colony of some 60 nests nearby. The new arrivals observe the locals' behaviour and go to find food at the dump. We mean, of course, the juveniles arriving for the first time, because the adults know where to go from previous experience. You start seeing juvenile west European birds in the middle of May, in small numbers, or rather proportions, because as the numbers of Iberian storks decrease those of west European ones increase. In the last week of August when the passage of Iberian storks is minimal the west Europeans are at their peak. Flocks are not as spectacular in numbers – you can count them in tens and not in hundreds like their Iberian counterparts in the Atlantic sector. These storks from outside Iberia mix with those Iberian storks at Los Barrios, which possibly come from eastern parts of Spain, and start the passage over the Strait together – although not all of them, as some groups spend practically the whole winter at this dump. Their migratory route leads them from the dump to the Tolmo-Faro area, in the Mediterranean sector, from where they cross over to Morocco.

Before finishing this chapter, I'd like to tell you about an experience I had at the Comisario lagoon during the postnuptial migration of the White Stork. In the month of June 2002 I was toying with an idea that I wasn't sure how to put into practice, but I was determined to go through with it no matter what. I had already taken photos of the tremendous concentrations of White Storks at the lagoon, but I wanted to take some from the same position as the storks themselves. There were two options: one was to place the camera in the roosting area with a wide-angled lens and operate it by remote control and the other was to stand in the middle of the flock and photograph whatever I wanted. The first option didn't convince me because I would have had to leave the camera for several

hours before the storks arrived and recover it late the following morning when the flock, or most of it, had left the lagoon. I liked the second idea but it seemed almost impossible to mingle unnoticed among several thousand storks without causing them to fly off in disarray…. To make a long story short, I decided to disguise myself as a stork. Yes, a White Stork! I asked my friend Antonio Luque if he could make me a mobile hide or stork disguise. He was delighted, if not somewhat surprised. During his spare time in July and part of August he put it together with copper tubing, wire and aluminium on the flat roof of his house. Gradually the bits of metal began to take shape and finally came to life. The result was a rather large White Stork (very well fed) and so heavy that some metal supports had to be fitted so that after walking for some time I could stop and rest. Inside it, apart from the photographer, there was a little platform where I could place a medium-sized 70-200 mm zoom lens, which was the main purpose of the exercise. On its first day out, my wife Palma del Valle, Antonio Luque and I took the stork – which looked like something out of the Valencian *fallas* – to the lagoon and positioned ourselves some 300 metres from a large group of White Storks, sunning themselves on a slope at the northern end of the lagoon. To make my disguise as realistic as possible I put on some red panty hose, exactly the same tone as storks' legs. Disguised and camera at the ready I set off, much to the amusement of Palma and Antonio, towards the peaceful group of unsuspecting birds. I tried to walk as slowly as possible and towards one side of the flock so as not to be too direct. As I got nearer some of them slowly took off, not visibly alarmed, but others followed. Gradually the enormous group got smaller and

smaller until there was only one left, standing with its back to me and apparently unafraid of my presence which had disturbed the rest. This cheered me up considerably as I thought that if one stork could tolerate my presence then so could a lot more and it would be just a case of improving my strategy. So, very slowly, I crept towards it on its right and from about four metres I took several photos. Suddenly I realized that there was something wrong with its right eye and it couldn't see me! I must have laughed and moved or something because the bird turned its head, saw me with its good eye and flew off startled. We had a good laugh about this for days afterwards.

A stage in the construction of the stork prototype and an action shot. Above, the one-eyed stork referred to in the text.

My friend Blanca
Román, ringing Stork
chicks at the river
Barbate dam in 2005.

The vultures that come from Africa every spring and pass over the Rock without stopping, return in autumn; they make their annual migration in flocks.

Francis Carter
A journey from Gibraltar to Malaga, 1777

Raptors and other migrants in the Strait

If the chapter on the migration of the White Stork has aroused in the reader what I intended from the very first line of the book, then I will have succeeded in making you leave your comfortable chair, turn down the music you were listening to and make a phone call to reserve a hotel room here in the Campo de Gibraltar, so as to share with those of us who live here the incomparable spectacle of flocks of hundreds of White Storks flying, against a background of bright blue sky, towards the Moroccan coastline. However, if you're still sitting comfortably, I suggest you don't put the book down but stay with me and familiarize yourself with the migration in the Strait of raptors and other birds, which will pass slowly through these pages, allowing us to observe their silhouettes while giving us looks that are always majestic, profound and defiant.

In the summer of 2003 I was at the Cazalla observatory together with several other ornithologists, when a flock of some 200 Black Kites appeared over us, heading for the coast. When they had passed over an experienced French ornithologist said to the rest of us: "Up to now I had only seen two Black Kites in my whole life and within a few seconds I've just seen two hundred all at once". Well, that's life in the Strait, a school for ornithologists, where in a few minutes you can see more species of raptors than you've seen in a lifetime of birdwatching in inland Spain or in Europe. Beginners can learn to identify raptors in a few days and quickly accumulate knowledge that would otherwise take them months or even years of work.

Do all raptors in Europe pass over the Strait of Gibraltar? Do they all fly in the same way? Are there any – like the White Stork – that

might spend the winter in Iberia? Is drifting an important factor for all raptors? We will be answering these questions and many more that will be raised in these pages, in as much detail as possible – within the scope of a book meant to inform as accurately as possible but without going into scientific details which might bore most readers – about the nature of the migration across the Strait of the different species of raptors, the Black Stork and some other interesting birds.

The status of local diurnal raptors

We think that this area is of considerable interest from the point of view of zoology, botany, scenery, cuisine – of course! – archaeology, music, the sea and so on; this is why in this book we touch on topics that the reader, at first, may not associate with the title on the front cover. A case in question could be the status of local diurnal raptors, which we mentioned in Chapter IV in migratory phenology under *Familiarity with local bird life,* and here we should like to build on that for the sake of those visitors who visit the area in the off-season (which is difficult) or wish to visit inland areas. In a region where so many different raptors coincide on migration it's possible that the observer may not be able to distinguish whether the bird he is watching is migrating or is a local one. Besides, when strong winds are blowing and migration comes to a partial or total halt, a visit to the interior, such as Los Alcornocales, affords one the opportunity to see local or resting birds and enjoy the interesting vegetation, especially during springtime.

Cádiz is a flourishing livestock-farming province and has one of the largest populations of Griffon Vultures in Iberia. In the area of the

Strait or, to be more precise, in the *sierras* and rocky outcrops of the Natural Park of Los Alcornocales, the most abundant raptor species is, without doubt, the Griffon Vulture, with over 800 breeding pairs. The most popular nesting area, of national importance too, is the Laja de Ciscar where in some seasons there are 90 occupied nests. In 2004 two deposits for dead cattle were set up in Los Alcornocales, one in the north and one in the south of the Park to offset the effect of the law which obliges farmers to bury cattle that die in the field. Luckily for the vultures nobody obeys the law, but when mad cow disease was a potential danger in the nineties the law was enforced with special vigilance and this reduced the number of dead cattle in the fields, thus threatening the species.

Another carcass-eating bird, the Egyptian Vulture, is probably the one that has suffered the greatest decline in the area over the last few years. There was an abundant population but now there has been a worrying drop. From the seventies, when it was common to lay down poisoned bait on all hunting estates, to date, the population has fallen so that it is one of the rarest birds in the area. On the La Almoraima estate, not one of the six pairs nesting in the seventies is left.

The population of Bonelli's Eagle seems to have remained stable over the past few years and this year a new pair has settled in an area where it was previously unknown.

The Peregrine Falcon is in a similar situation and doesn't seem to be on the decline, although the pairs we knew which nested on fairly low rocks have changed their nesting sites to much higher, inaccessible places. In the early eighties there was a nest where, if you wanted to photo-

Opposite

Two Griffon Vultures near Torre del Fraile, about to reach the European side of the Strait on a day of light poniente winds.

graph the chicks, you had to squat; it was in a rock cavity only 1.5 metres above the ground.

Forest raptors – and by that I mean those that breed in or on the edge of forests – may experience survival problems because their territories are "invaded" by cork harvesters every nine years. No study has been made on the impact on raptor populations in Los Alcornocales caused by these groups of cork cutters who spend two or three days there and it could well be that due to their presence many adults abandon their chicks or the chicks die for lack of food. The harvest starts in the first week of June, coinciding with the hatching of Short-toed Eagle, Booted Eagle and Goshawk chicks. We mustn't forget either that gangs of men prepare the ground some days in advance so that the cork oak trees are accessible by harvesters and mules which have to transport the cork to clearings and all this means noise and disturbance. It used to be quite common for the men to take a chick from its nest and feed it during the cork season, never to be heard of again. I have met harvesters with chicks of Short-toed and Booted Eagles, Goshawks and Little Owls etc; nowadays it's not a common practice. As far as Goshawks and Sparrowhawks are concerned it is very difficult to find their nests and I wouldn't like to speculate on how they are faring, but I would bet that the nests I found many years ago are still there, although I am not aware of any new ones.

The Booted Eagle together with the Short-toed Eagle is the most abundant raptor in the area, certainly much more so than the Goshawk or Sparrowhawk. All the nests I have seen, and I've seen a lot, have contained two chicks and all the adults were pale phase birds.

Although I hardly ever manage to see a snake, lizard or skink when I'm out in the coun-

Goshawk, with a single chick, alone in the impenetrable solitude of a deep forest of mixed Portuguese Oak and Cork Oak trees, in Los Alcornocales Natural Park.

The Short-toed Eagle nests in cork oak forests near flat pasture land.

try, the Short-toed Eagle nests in practically every wood in the area, always on a side branch of a cork oak at the edge of the wood and the scrub. I know of no Short-toed Eagle's nest on a Portuguese Oak probably because these grow in shady areas and this eagle likes sunny areas. All the nests are built on small cork oaks and in them I've often found remains of disembowelled toads, vipers and even cattle egrets.

On one occasion, José Rafael Garrido - now a biology teacher at the Pablo Olavide University in Sevilla – and I were at the observatory at La Peña, while studying the migration of the White Stork, when we saw a local Short-toed Eagle fly past the foot of the eastern side of the *sierra* de Enmedio, quite near us. A flock of Starlings flew in the same direction a short while later and, while we were watching a flock of storks gliding towards Tarifa Island, we heard a strange squawking noise. Glancing in that direction we saw a Short-toed Eagle with a starling in its claws!

Paradoxically, the Black Kite is one of the rarest birds in the area. I know of only five nests in the whole territory and that in spite of several thousand of them passing this way every year and the existence of a cork oak wood at the Los Barrios rubbish dump which would be a good nesting place for them, only a few metres from a fast food "restaurant" for them and their chicks.

The Buzzard is less common than the Booted Eagle, and there are fewer of them in the south than in the north of Los Alcornocales.

The Honey Buzzard and the Hobby are passage birds; the Red Kite is only a winter visitor; the Osprey is also a wintering bird with more than half a dozen individuals on rivers and reservoirs; the Black-winged Kite nests very near the area and is a rare passage bird. As for kestrels the Common or Eurasian Kestrel is abundant but the Lesser Kestrel has seen an inexplicable and alarming decline in its population both in this area and in general in the province of Cádiz during the last five years. Back in 1994 I counted 31 nests in the centre of the city of Algeciras and now there are only 3 left. A similar fate has beset the colonies in Castellar, Gibraltar and Los Barrios.

The Black Stork

The enthusiasm for this migrant existing in the area never ceases to surprise us, not because we don't share the admiration for this wonderful bird, but because we don't understand why – with a passion almost comparable to that of a football fan – many aficionados only come to the Strait to watch Black Storks migrating and hardly raise an eyebrow at flocks of White Storks

in much more spectacular numbers and much easier to see. In September when many Spanish and foreign ornithologists come down to watch the migration and a flock of Black Storks appears, no matter how small, a buzz of excitement comes from their fans, whose indolence is transformed into feverish activity: they make a charge for the telescopes, other ornithologists retire so as not to get in their way and you can hear shouts in several different languages. It is not for nothing that one of the most important local NGO's bears the bird's name: the Black Stork Ornithological Group (COCN).

The Black Stork (*Ciconia nigra*) is a monotypical species (ie it has no subspecies) found in the Palaearctic and winters in Northern India, north-east Africa, south-west to east China and in the south-west of peninsular Spain. A few pairs also breed from Malawi and Namibia down to South Africa (Del Hoyo *et al*, 1992). In Iberia they are found mainly towards the south-west, especially in Extremadura, Castilla-León, Andalucía, Castilla-La Mancha and Madrid, nesting preferably on rocks, in pine and cork oak woods, and always near water courses and in scarcely populated areas. The number of known breeding pairs is currently 387, of which 173 are in Extremadura (Martí, R. & Del Moral, J.C. Eds, 2003 *Atlas de las Aves reproductoras de España*). In our opinion it doesn't breed near the Strait, possibly because the cork oak forests are on steep ground; there are practically no cork oak *dehesas* here as there are in Extremadura. Nor are there any wide rivers because these rise very near the coast and reservoirs are not shallow at the edges so the birds cannot wade. What we do have in Los Alcornocales is that other pre-condition for Black Storks, rare human presence, but the land is very hilly and the woodland is dense.

For three years from 1998 to 2001 a study was conducted in a part of the National Park of Doñana, where 54 to 58 Black Storks from Iberia and central Europe spent the winter.

Although a trans-Saharan migrant, many individuals spend the winter in central and south-west Iberia, just like the White Stork. Those that winter in Africa arrive on the European side on prenuptial passage in March and April. I have seen them arrive in the Natural Park of the Strait but never in large flocks. The largest I've seen had twenty or so individuals. Because prenuptial migration takes place mainly in March it is not well recorded because the weather is often very wet, with day after day of strong winds. There are much better records of postnuptial migration, as we shall now see.

As far as routes are concerned, the European migration of this ciconiform is very similar to that of the White Stork on postnuptial passage: the eastern route over the Bosphorus and Dardanelles, with over 10,000 individuals; the Strait of Messina in the central part of the Mediterranean, where in 1999 139 Black Storks were counted; and the western route via the Strait of Gibraltar. The latter is the route for Iberian Black Storks and those from central Europe, which reach the Peninsular via either end of the Pyrenees.

The first studies in the Strait were conducted in the seventies, led by Bernis, and yielded 373 birds crossing in 1976 and 191 in 1977. In 1993 César Sansegundo published the results of studies we mentioned earlier on: *Recuento de cigüeñas negras en migración otoñal por Gibraltar* (Quercus nº 102:13-16). In 1995 a study group was formed, mainly made up of ornithologists from the Campo de Gibraltar area and they did great work with a census of birds crossing from

The Black Stork is an extremely elusive bird and avoids proximity to humans, but the Iberian population is increasing very slowly.

that year until 1997, publishing the results on the Internet under: *Censo migratorio postnupcial de la cigüeña negra (Ciconia nigra) en el Estrecho de Gibraltar*. This census was carried out by Cristina Parkes, Manuel Lobón and J.M.Jiménez as part of a wider study conducted by the Black Stork Ornithological Group (COCN) and in 1997 out of a total 3,512 birds, 2,296 birds crossed the Strait. The Migres Programme also monitored this migrant bird on postnuptial passage in the same year and two things became clear: one, the split between local ornithologists and the Migres Programme as a result of the lamentable policy of completely ignoring those who, well before official bodies became involved in Migration in the Strait of Gibraltar, were already gathering and publishing information; and two, the duplication of studies on the same species. While the locals counted 2,296 birds crossing, Migres came up with 1,469, i.e. 36% fewer. Regardless of the fact that these data are completely different, it is true that numbers crossing the Strait of Gibraltar have gone up considerably, as have the populations on the other side of the Pyrenees and in Iberia, which is good news at a time when population data always seem to be negative.

Birds approaching the area of the Strait come from three directions: the west, via the Valle del Santuario and river Valle valley – in this case just as in the migration of the White Stork, sometimes on a N-S course, others on a NE-SW; the east, when the birds arrive at the Sierra del Algarrobo near the city of Algeciras, also on a NE-SW course; and, finally, via the Rock of Gibraltar. The latter course means they either cross the Bay of Gibraltar heading for Alelíes in the Natural Park of the Strait, or, on days when the wind is in their favour (and even when there

are strong crosswinds), they start the crossing immediately as they are strong fliers. According to Bernis (1980) the Black Stork doesn't cross with moderate *levante* winds, only when these are light or there is a westerly, but COCN concluded that they cross with a prevailing *levante* of force 7 on the Beaufort scale, whereas the Migres Programme reported that only 5% crossed when this moderate wind was blowing.

The largest recorded flock, made up of 676 individuals, was seen on 30th September 1997. (This figure was unthinkable in the seventies when Bernis counted a total of 373 in 1972) On that day a total of 731 individuals crossed the Strait and the previous day 246. These crossings took place after the first autumnal rains in the area. At the observatory in Gibraltar set up by the GONHS at the cable car station 328 were counted, a record number for Gibraltar.

In the migration of the Black Stork there are certain differences when compared to that of the White Stork. If the flocks are of a reasonable size, the cohesion among its members is there but is less obvious because they fly at a much higher altitude than the White Storks. Their migratory behaviour as regards age is, moreover, not well documented. It doesn't seem that the juveniles migrate first, followed by the adults, although flocks of juveniles only and adults only may be observed. We have already mentioned how flocks of White Storks, which fail to get past the coastline at their first attempt, regroup and then try again. Black Storks, however, are seen wandering along the coast making no apparent attempt to cross again or they turn back and head northwards, disappearing from view. This behaviour was well documented by COCN in 1997 when, out of a total of 3,512 birds observed, 2,296 crossed the Strait, 473

were "wanderers" and 743 returned northwards after reaching the coast.

From mid-September until the first week of October is when 90% of Black Storks cross the Strait, concentrated almost exclusively between 7:00 a.m. and 15:00 p.m. No crossings were detected before 7:00 a.m. or after 17:00 p.m.

On several occasions a Black Stork has been spotted in a flock of White Storks as well as one, or two, of each species flying together. As you can see from the following anecdote, this species can be quite sociable.

Flock of Black Storks, in migratory formation, gliding at medium altitude against a light levante wind near Tarifa.

In 2002 my friend Antonio Luque and I set up a hide so as to be able to photograph the gatherings of White Storks at the Comisario lagoon. In the afternoon hundreds of storks started landing at quite a distance from where we were hidden, when, while watching the "show" one us spotted two Black Storks flying directly above the colourful group of thousands of White Storks. They gradually floated downwards to try and land among this huge flock, but when they tried to touch down in the shallow water the nearest White Storks hit out at them with strenuous wing flapping. The couple tried to land again and again but each time they were beaten off and eventually flew out of our sight. This reminds me of the number of times I have tried to locate their roosting sites. I have heard of Black Storks resting and spending the night in Nerja, near the sea (Antonio Jiménez, personal communication, 2003), on the El Jautor estate (Alcalá de los Gazules) and near the Almodóvar reservoir (Tarifa), but there don't seem to be any traditional roosting sites near the Strait. This species just stops whenever it comes upon a quiet place to rest.

One of the latest studies which monitored the Black Stork, in fact the most popular one there has ever been, was called Africka Odysea, promoted by Czech Public Radio. Monitoring the Black Storks of this country enabled them to reveal the migratory routes they follow and the dangers they face on the way, both in Europe and in Africa. Satellite transmitters were attached to ten birds and their routes to Africa were monitored, some of them via Gibraltar and some via the Bosphorus. A male, named David, spent the winter of 1998 in Andalusia; Zuzana was killed in Ethiopia; Vaclav flew from Bohemia to the Balkans and from there to the Ap-

The Honey Buzzard is one of the star raptors migrating via the Strait of Gibraltar. Birdwatchers marvel at the compact flocks coming over in large numbers.

penines where he was shot down, as was Hynek in France and Jonas in Spain in 1999. But the star performer of the project was Kristyna, followed each year for four years from the Czech Republic to Senegal, where it wintered. All the information is available on the Internet at Africkaodysea@internet.cz

The Honey Buzzard

At the end of the sixties on a misty May morning I was resting on a small beach called Arenillas in the Strait of Gibraltar after a tiring session of deep sea fishing, when I saw a group of ten raptors flying low, possibly because of the thick mist, and slowly soaring upwards by a small cliff. I grabbed my 8x30 binoculars, a present from my father, and watched them excitedly, trying to memorize as many details as possible to be able to identify them. When they disappeared into

Honey Buzzard, one of a compact group out of shot, gliding towards the Strait coastline.

the mist shortly afterwards I frantically flipped through the pages of my "Peterson", full of pictures of birds in flight or at rest. After examining those in flight closely I stopped at one: the Goshawk. I was sure I had seen Goshawks because of their wide wings, long striped tail and their unmistakeable silhouette. I had seen ten Goshawks just when they had reached the European coast of the Strait, after flying probably several thousand kilometres over savannahs, dense forests and bone-dry deserts. Later, back at home I got out my migration form and filled in the details of my fantastic discovery. Some years later I found out that when my form arrived at the SEO (Spanish Ornithological Society) there were some amused comments in the Bird Migration Section wondering who on earth this Fernando Barrios was who had seen no fewer than ten Goshawks migrating together. After a time I learned that the Goshawks were in fact Honey Buzzards and all I had noted down were the wide wings and long tails!

The Honey Buzzard (*Pernis apivorus*) is a monotypical species widely distributed in the western and central Palearctic. With a European population of an estimated 160,000 breeding pairs its presence in the Mediterranean peninsulas (Iberian, Italian and Greek) is scarce (Del Hoyo *et al.*, 1994) and it winters in sub-Saharan Africa, preferably south of the Equator. In Spain it breeds from Galicia to Navarra and in the Pyrenees as far as Gerona. There is a major nucleus in central Spain and in the provinces of Cáceres and Badajoz. In Iberia as a whole there are about 1,150 breeding pairs of which about 1,000 are in Spain (R.Martí *et al.*, 2003). A notable exception to these breeding sites is one recorded in Málaga in 1999 (Pons *et al.*, 2000). The diet of the Honey Buzzard is made up mainly of in-

sects, preferably wasps, although it is also known to capture amphibians, reptiles and small birds (Castroviejo and Fernández, 1986).

Compared to other migrant raptors and soaring birds in general, that cross the Strait of Gibraltar, the Honey Buzzard has several distinguishing features. The first one – and a very attractive one it is for bird watchers – is the different type of plumage in the species, ranging from black to white with different intermediate grey tones; the second is that the time scale of its passage is very short compared to that of other raptors; the third is that they fly in large flocks; and the fourth, and considering that the majority of Honey Buzzards arrive in the first week of May and return in the first week of September, is that they have little time to rear their chicks, the success rate being 33%. These are the main distinguishing features, but there are others, as we shall see.

In his *La migración de las aves en el Estrecho de Gibraltar. I Aves planeadoras* (1980) Bernis says, when referring to the migration of Honey Buzzards in the summer/autumn of 1972, " Their impressive passage *en masse* via Gibraltar, seen then for the first time in its true dimension, took us completely by surprise". It certainly surprised me the first time I watched the passage of thousands of Honey Buzzards one day in early May in the mid seventies at a petrochemical factory where I was working at the time. I recall it was a day with a light wind, possibly a westerly. A colleague of mine was fascinated, saying he had never seen so many eagles together. I thought he must be talking about a flock of vultures which patrolled the area regularly and that, for him at least, more than a dozen was a "terribly large" flock, as they say in our area. It must have been about 8 in the morning and I decided to go with him to have a look at these "vultures". We went to a part of the factory where there weren't many pipes or pumps or distilling towers and, quite frankly, I was flabbergasted. Several hundred Honey Buzzards were flying over the factory, some of them gently flapping their wings, their plumage, ranging from white to almost black, standing out in stark contrast to the intense blue sky. From where I was standing the whole sky was covered in Honey Buzzards, flying in a huge, compact flock. Judging their altitude by one of the 40 m high distillation towers I reckoned they were flying at a height of 30 – 45 metres. It was one of those unforgettable scenes, rarely witnessed and which remains engraved in your memory; even though it was a long time ago I can still close my eyes and see hundreds of birds flying, as if performing a ballet especially for us. For many, many minutes – I didn't time them – they were passing overhead and many of my work colleagues came out to join us and watch, absolutely captivated by this endless procession of raptors.

In Chapter IV we said that soaring birds, as a general rule, funnel towards straits on their pre- and postnuptial passages. But this doesn't always apply to the Honey Buzzard. From the seventies, when studies in the Strait began, until the present day, little or nothing has changed in the way birds migrating across it are counted, although there are other important methods of gathering and divulging information and knowing the routes migrants follow. Today many more people are interested in migration and observe it in person and then publish data on the Internet, where you can find any forum imaginable. Nature is in vogue and for some it is even a way of life advocated by the media, so that nowadays many enthusiasts covering a very large area have become observers, some of whom are very well and some not so well informed. The improvement in the economy over the last few decades is a major factor in the increase in public interest in a pastime, which was only open to those who were relatively well off or who had an almost fundamentalist and exclusive passion for birdwatching. The scientific ringing of birds and, especially, the use of satellite-linked radio transmitters have been extremely helpful in the monitoring of large and medium-sized birds on their migratory routes with an accuracy of 15 metres. When discussing migration now, it is possible to learn about the passage of birds in hitherto unknown places, because before the amount of human resources necessary would have made the cost prohibitive.

This preamble is to introduce the latest knowledge we have obtained from the satellite tracking of Honey Buzzards and from the large number of people observing these events in nature and posting comments on the Net – although of course these must be scrupulously analysed, contrasted and verified accordingly. What we mean by this is that not everything you read on the Net can be accepted, because there are people who, wishing to please, will give information that is not real or is, at least, an exaggeration. But, if you are patient, you can generally spot this kind of behaviour and ignore it.

So, thanks to the increase in the number of observers and to satellite tracking of Honey Buzzards, we have been able to discover things about them, which would have been unthinkable years ago. Until the eighties it was thought that the migration of the Honey Buzzard was channelled exclusively through the straits in the Mediterranean. They have always been classed as excellent flying birds, but not to the same degree as falcons

Flock of Honey Buzzards in a thermal.
These flocks can vary from just a few to
several hundred.

and harriers, which do not depend on the straits when crossing the Mediterranean. In southeast Sweden satellite transmitters were attached to 6 adults and three chicks and the surprising news was that the adults crossed over to Africa via Gibraltar, continuing over the Sahara desert to Sierra Leone in West Africa and from there on to Cameroon where they spent the winter. From these data you can see that these adults went on a big detour to get to the wintering site. The juveniles, on the other hand, flew due south, over the Mediterranean with no detour and wintered in the same place as the adults. Cruising speed was similar in both adult and juvenile, 170 km/day in Europe, 270 km/day over the desert and 125 km/day in sub-Saharan Africa. The adults rested less time (42 days) during the journey than the juveniles (64 days). This study states that age is a very important, segregationist fact in migration because, on postnuptial passage the juveniles follow a route which doesn't include crossing straits and on prenuptial they are dragged into the migratory current and cross them all together. However, in 2004 we were witnesses to something very unusual and which puts into perspective the claim that all Honey Buzzards return via Gibraltar.

For years in the Balearic Islands when there is a prevailing *poniente* wind a few Honey Buzzards have been detected passing over, but it was always thought that this was due to a drift effect and only the odd group of birds used this route. This was also partially confirmed by results obtained from satellite tracking of birds from Scotland and Sweden. It is currently considered to be a scarce to moderate migrant (GOB, 1999). At 42.4%, and less than 100 individuals, it is the most common raptor species on prenuptial passage over the Cabrera Archipelago (*Migración e*

invernada de rapaces diurnas y nocturnas en Cabrera, Rebassa, M., 2000) In the rest of the Balearics nearly all observations have been on prenuptial passage, with a maximum of 70 birds in Formentera (*Anuari Ornitològic de les Baleares,* GOB, 1995)

But in the month of May, 2004, thousands of Honey Buzzards were detected passing over the Balearic Islands, with prevailing northwesterly winds, which goes against the previous theory and presents us with the Honey Buzzard as a migrant bird that does prefer to use straits but, when necessary, is not at all afraid of flying over large tracts of water. Therefore, the approximately 100,000 individuals which cross via Gibraltar on postnuptial (there are no figures for prenuptial) migration are only a fraction of those that migrate to their wintering areas in Africa. Those detected in the Balearics were flying during the middle of the day, which means that they will have left the coast in the early morning to

Route normally taken by Honey Buzzards across the Strait :

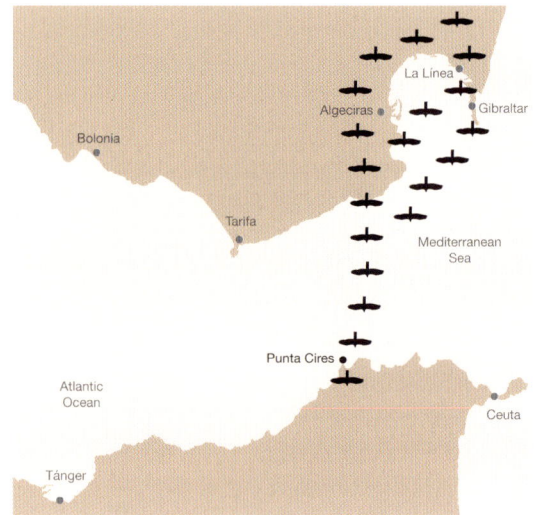

make a non-stop crossing of the Mediterranean (Maties Rebassa, 2000)

If we look at how the Honey Buzzard behaves on prenuptial migration we see that this starts in April and, as with nearly all migrant species, during the first few days you observe the odd individual. These are followed by several days of small groups of a dozen or so, and then suddenly flocks composed of hundreds and immediately after you get the huge migration of thousands of birds every day. The flocks are not as compact as those of the White Stork; they are more in the style of the Black Kite, which fly together but need *breathing space* between individuals. Perhaps one notable difference, when compared to the migratory timetable of other raptors, is that they don't always need thermals to start crossing. You can see this quite clearly on prenuptial passage because at first light some flocks are already arriving on the European side of the Strait: on postnuptial you also see flocks flying to Morocco at dawn. If a bird watcher is not an early riser and thinks that at 8 a.m. or so it's not worth going to the Strait, then they will miss the main body of these migrants.

Thermals are not as essential for these powerful birds as they are, for example, for Griffon Vultures. Depending on the winds you can see compact flocks arriving just after dawn, usually when it's a westerly, because this particular wind drops in the evening and at night and only gathers strength as the day advances. Perhaps the Honey Buzzard takes advantage of this and crosses the sea when winds are less violent and navigation is easier.

On a day of heavy passage traffic, rather like the rush hour in urban life, more than 10,000 birds may cross. This colossal number of migrants assembled together is one of the great zo-ological phenomena and a sight of great beauty. Ornithologists or not, few people can fail to be moved by this. Such a huge gathering, however, is not just one flock, but various, quite large flocks, which come out of the Moroccan hinterland almost continuously. Probably the ones crossing at first light are those that have spent the night near the Strait and they are followed by ones from further away. When evening comes and the effect of thermals is minimal, they will stop and rest until the next morning.

Many observers have seen flocks arriving at dawn, when the sun has just appeared and there are others, like Luis Barros of the Migres Programme, who has told me of a time when in the month of May, with night vision binoculars he was using for another job, he saw a flock of Honey Buzzards arriving at 9 in the evening near one of the wind farms, fly overhead and on, probably looking for a good place to spend the night.

Drift caused by strong crosswinds affects Honey Buzzards, but possibly not as much as it does other raptors, because they are gifted with powerful wings and are not afraid of drift. They have been seen to cross even with storm force winds, so it isn't surprising to see them drift as far as Malaga or even further east with a strong westerly and as far as Conil and further west with a strong *levante* wind. The doubt arises, when it comes to drift, whether these birds allow themselves to be dragged as far as the Balearics, not because they start off near the Strait, but because they leave from some point east of Melilla or even Algeria and because of either *poniente* or *levante* winds they drift so much that they have to pass over the Balearic Islands. With *poniente* winds it is quite feasible for the North African front to stretch from Tangier to the

Algerian coast and beyond that point there are flocks which cross via the Strait of Messina. The passage of thousands of Honey Buzzards over the Balearic Islands, then, cannot be put down solely to the drift effect of winds on individuals from the area of the Strait of Gibraltar.

As the flocks are so large and, during the migration season, they are constantly flying over the observer who is lucky enough to be in the right place at the right time, a special situation arises, which we shall try to explain as precisely as possible to the reader.

Let's imagine that a flock of Honey Buzzards leaves the Moroccan coast at first light and the prevailing wind is a westerly force 1 on the Beaufort scale. This means the crossing route will be in a south-north direction. If, as the hours tick away, the wind reaches force 3, drift will take them a little further east, let's say as far as the Tolmo inlet. Time passes and the wind increases to force 4, so that the flocks are now passing on a front over the Punta Carnero lighthouse; with a force 5-6 they're crossing over Gibraltar and, as the wind increases more and more they are now crossing over Malaga and beyond.

What would happen if an observer stayed put all this time at the Punta Carnero lighthouse? They would see the first flocks of Honey Buzzards approaching west of their position, over Punta del Fraile and gradually coming closer and closer until, feeling really excited, they would see the Honey Buzzard *armada* right overhead. Later, as the wind gets stronger they would watch them going further and further away until they disappear from sight and they would think the birds have stopped crossing, when in fact hundreds of them are still doing so but much further to the east and out of their binoculars' vision.

So, an ornithologist not familiar with the Strait would think they had seen hundreds of Honey Buzzards for a good length of time, many of which flew right above their head, only to see them slowly disappear and then believe that no more crossed the Strait that day. But if the wind strength stays the same and, on not seeing any more birds, our ornithologist decides to drive off to Gibraltar and do some shopping, when they get there and look up – as birders do constantly – they will see more flocks flying over the Rock and be convinced that Honey Buzzards have started crossing again, when it is in fact the drift which has carried them in that direction.

To give an even more graphical description, the reader at home should get out a map of the Strait and place a ruler in a north-south position, the southern point being fixed at Punta Cires on the Moroccan side and then slowly move the ruler's northern end. The sector covered by the moving end of the ruler will be that covered by the fronts of Honey Buzzard flocks. As the wind strength increases, when blowing from the west, the directions of the arriving flocks will begin to move eastwards, starting at Tolmo-Faro, then moving to the Bay of Gibraltar, then the Rock itself and so on until they reach Málaga and beyond. Logically with a *levante* wind the same will happen but towards the Atlantic sector of the Park. These comments about the Honey Buzzard are also applicable to all raptors, but in this species the movement is more noticeable because it flies in large flocks and against strong winds. Griffon Vultures also fly in large numbers, especially over the last ten years, but they need different wind conditions to those of other raptors, as we shall see shortly.

The altitudes they arrive at on the European coast of the Strait may be medium, high or very high. As with the White Stork, rarely, if ever, do they arrive at the coast flying lower than 50 metres. In all our years of ornithological and photographic experience we have only witnessed this on two occasions when there was a thick mist, but never because of a strong wind. These birds flap their wings strongly over long distances and never look visibly tired.

In 1978 I was fortunate to see the passage of Honey Buzzards over the Moroccan desert, near the village of Merzuga. They were flying low, about 30 metres high, flapping their wings constantly. I saw several groups of between 80 and 200 birds each and they didn't make use of thermals. They just flapped their wings as I watched them, I don't remember for how long, but it was several minutes before they disappeared from view.

When they get to the European coast they start gaining height in spectacular thermals containing hundreds of birds, then head north out of sight like all other migrants, with the exception of those like Griffon Vultures or Short-toed Eagles that land on the ground to rest.

On their postnuptial migration the majority of birds that arrive in the Campo de Gibraltar area come from Scandinavia and west-central Europe, flying generally across inland Iberia; this is backed up by the fact that there are few reports of sightings of Honey Buzzards at this time down the Valencian coast, whereas there are many in that area on prenuptial migration. Bernis reports that they easily overcome mountainous areas - 3,200 metres above sea level is no problem at all for these magnificent flying machines! On prenuptial passage they are almost bound to cross the Atlas Mountains in Morocco from south to north without having to stray from their route to get over these great heights.

Nevertheless, sometimes it may be hard to be able to see some flocks under certain weather conditions: when the winds in the Strait are very light and on those rare days when there is an absence of crosswinds, inexperienced observers may not see a single Honey Buzzard for hours or in a whole day and what is happening -and it often happens - is that the great majority of flocks are flying over their position at a very high altitude, so high that even with 10 power binoculars they are just little dots in the sky and extremely difficult to detect. That is why you must be extremely alert when watching soaring birds in the Strait and to spot Honey Buzzards you really must keep your eyes peeled! The flocks come down from the interior beating their wings intensely and, quite determinedly, they disappear, Morocco-bound. It is neck-aching work for the serious ornithologist who doesn't want to miss a flock out of the day's count.

I have nothing but admiration for these superb birds, masters of migration; they have given us many unforgettable moments and enjoyable days in the Strait as we watch them passing nonstop throughout the day, before returning home after a memorable day in the field. They are the reason why, in the month of September when the greatest number of ornithologists in the observatories coincides with the appearance of these huge flocks of Honey Buzzards, the smiles of satisfaction last throughout the hot and / or windy day.

In the chapter on the White Stork we said that 98% of the Iberian storks arrive at the Coast on the Atlantic side of the Strait and almost 100% of those from outside Iberia arrive on the Mediterranean side; we also mentioned the influence of drift which causes the birds to gather in areas opposed to the direction from which the wind is blowing (with a *levante* wind birds tend to gather in the Atlantic sector and with a *poniente* in the Mediterranean sector). However, most Honey Buzzards come to the area from the north-east, possibly following the route that took them to the breeding areas of central Europe and Scandinavia, just as the non-Iberian storks do. When there are westerlies they come to the Strait coast from the east, via Málaga, following the coastline or flying parallel to it just a bit inland. A strong east wind will bring flocks along the coast or parallel to it but in the Atlantic sector or from the interior towards the Los Lances area and they will then cross at Alelíes. On extremely violent windy days you can see flocks of Honey Buzzards advancing skimming over the meadows, while the rest of the migrants, like Black Kites and White Storks, are sitting in trees or on the ground. And they will still arrive from the east with prevailing strong easterlies without going over Gibraltar. They come down from the hinterland arriving at Algarrobo from the northeast, and there skirting along the *sierra* they turn southwest or south-southwest and fly towards Tolmo-Faro or Alelíes and cross from there.

There are many days when, with a prevailing westerly, you can see Honey Buzzards flying over La Línea de la Concepción and the Rock of Gibraltar. From Upper Rock you can see them coming from the interior, from the *sierra* Crestellina (Málaga) and approaching the coast, then usually going round the east or west side of the Rock. In the early morning these flocks fly low over the buildings in the city and, as the temperature rises, they fly higher and higher. It is a terrific sight from Upper Rock, watching them fly over the Rock and La Línea. Depending on the time of day and the prevailing wind many flocks cross towards Morocco from the Rock itself or cut across the Bay and start their crossing from Tolmo-Faro or Alelíes.

On certain days some migratory "corridors" are established and all flocks of Honey Buzzards crossing the Strait go along them. You see a flock coming and, when another appears half an hour later, it will go along the very same route as the one before. The same occurs – and we forgot to mention this – with the White Storks. These migratory corridors, once detected, make life a lot easier for observers, because, besides trying to spot flocks all over the area, they keep looking at this corridor and see flocks they might otherwise have missed. Sometimes, however, these corridors change places, so that when a flock passes over a different place from the one being monitored, the following flocks also do the same. It is as if an area existed where, because of the terrain or the type of thermal, a row of thermals is formed and the birds use them up one by one.

Another characteristic of Honey Buzzards is that they are much less affected by drift than other raptors, despite being blown way off course. This is more evident on prenuptial passage because you can compare them with Black Kites at the same time. You can see Black Kites flying near the coast only one or two metres above the sea, sprayed by the waves. Although the wind is less strong at that level they still drift considerably, whereas the Honey Buzzards can fly at medium altitude and not budge from their chosen course. This doesn't mean they aren't affected by strong winds, but we just want to point out that, while for some migrants a given wind strength will test them to the limit, that same wind will pose no difficulty for the Honey Buzzard. As they are very powerful flyers they don't hesitate at tackling the Strait crossing under storm conditions

A Honey Buzzard on its Strait passage,
has stopped to dig out a wasps' nest.
Such behaviour is unusual while migrating.

(force 10 on the Beaufort scale), although they are subjected to considerable drift, which will take them beyond Cádiz or Málaga. We think, however, that when the winds are not too strong and you see Honey Buzzards over Murcia and Alicante in the eastern Mediterranean, then they are either birds that have flown east along the coast from the Strait or they have crossed the Mediterranean south to north but not via Gibraltar. Logically, such cases are difficult to detect if an observer isn't on the coast watching them arrive over the sea. We are certain that many Honey Buzzards counted as incoming migrants over those coasts are not new arrivals, but have really arrived via Gibraltar.

During the studies conducted by Bernis in 1976 our friend Cristina Carro spent a day in Ceuta watching the passage of Honey Buzzards with a fairly strong to strong *levante* wind depending on the time of day and she counted 3,000 of them, of which 2,500 arrived in just one hour. According to Bernis all the birds passed over the stretch between Monte Hacho and the border with Morocco at Benzú. He came to the conclusion that the flocks had drifted 10 to 12 km to the east. This was the first time that the effect of crosswinds on particularly skilled flyers like Honey Buzzards had been recorded – and if you think of the dramatic effect this drift would have on Griffon Vultures, then you realize why they only cross on days when there is a light or moderate wind.

Another similarity between Honey Buzzards and White Storks is that winds don't influence their choice of where to cross the Strait: the majority of them cross via the Mediterranean sector and Gibraltar. Very few cross over Los Lances and sometimes they take a southwesterly course from the Rock and at the Algarrobo observatory (Huerta de Serafín) you can see them flying along the coastline over the beach at Getares and on towards Tolmo-Faro, or a bit further on to Alelíes, from where they will cross the Strait.

Again, we must ask ourselves if Honey Buzzards only cross over to Africa via the Strait of Gibraltar. Thanks to the latest developments in monitoring birds by satellite tracking their exact migratory routes have been discovered – ringing doesn't give you this information. A Honey Buzzard ringed in Scotland and captured in Mali would suggest it crossed the English Channel and then crossed to Africa via Gibraltar, but nothing else. If we look at the studies conducted by the Highland Foundation for Wildlife which satellite-tracked some adult and juvenile birds, we will see that our suppositions were not as accurate as we thought.

In 2001 transmitters were attached to two chicks a few days before flying. One of them, number 21252, left the North of Scotland on 7th September. Its signal was lost before crossing the English Channel but fortunately it was recovered near Cartagena (Murcia) on 23rd October and it was from there that it crossed the Mediterranean to the Moroccan-Algerian border. On the 27th the signal was lost completely a bit further to the west. The other chick, number 21253, left Scotland a day later, but its signal was lost somewhere over the Atlantic.

So, after discovering that Honey Buzzards cross via the Balearic Isles on prenuptial migration and those coming from Scotland on postnuptial, we can be sure that their flyways are via France, Catalonia, the Levante etc., but the number of those crossing increases the nearer they get to the Strait.

Unlike most soaring birds which have to wait until the sun warms up the land so that thermals start forming, the Honey Buzzard isn't totally dependent on this form of "transport" and can do without them. The irrefutable proof of this is that they have been satellite tracked across the middle of the Mediterranean Sea. We have seen flocks of Honey Buzzards approach the Algarrobo observatory at first light, when the sun is just a red disc and it is still quite chilly. They will probably have spent the night in the woods near the *sierra* and left them at sunup. These first flocks fly low, but other flocks start gaining height to migrate when the sun warms up the fields and thermals begin to form. Most migratory activity in the Strait stops around 17:00 p.m. but it is not unusual to see flocks of Honey Buzzards crossing until it is nearly dark, although not at the same rate as during the day.

Many migrant birds eat while migrating. Some, such as Swallows and Swifts, eat insects as they fly and don't have to stop as they are continually refuelling in mid-air. White Storks, Black Kites and Griffon Vultures refuel too, but at rubbish dumps, where their migrating instinct is temporarily replaced by a need to eat. The great majority of Honey Buzzards, on the other hand, eat absolutely nothing until they reach their final destination, although, like everything in life, the exception proves the rule.

One year during the month of May I was driving along a path with a warden inside the magnificent La Almoraima estate when, on turning a sharp corner, we saw on the driver's left a raptor on the ground, flapping its wings. When it saw us, it kept still for quite some time and then, as it took off, I realized it was a Honey Buzzard. We got out of the old Land Rover and saw that the bird had made a hole in the ground, apparently "looking for wasps' nests" according to José Luis Márquez, the warden. He said it was unusual to see such a raptor and only the second time he had seen one, both times looking for wasps.

An even more surprising anecdote came from Joaquín Mazón, an ornithologist working on the Migres programme, who told me recently that in September 2005, near the site of the old Janda lagoon, he saw some Honey Buzzards roosting, when suddenly one of them flew down from a branch near the road and started digging in the ground to get at a wasps' nest. He was able to get as close as five metres from the bird and photograph it, feeding, for a few minutes without it flying off.

How many Honey Buzzards pass over Gibraltar and other major passage places in the Mediterranean? This is a difficult question to answer because the Honey Buzzard is a bird that dares to cross in weather conditions in which other birds would probably find it impossible even to fly. Thus, on very windy days when it is quite a sacrifice for observers to stay at their posts, if they have to be on the lookout for Honey Buzzards for the census it can be an arduous task sticking at it for hours on end. Besides, there are some flocks that fly *stratospherically* and even the best ornithologists in the world are unable to detect them every time. And to make things even more difficult, when what you are trying to achieve is a rigorous count, their passage timetable is different from that of other raptors and soaring birds. They sometimes fly to Morocco at dawn or arrive on the European coast on prenuptial passage at first light. As for when migration stops, well, we have already seen that flocks coming over from Africa have been detected at night, so, all in all, counting these migrants when they pass over the Strait of Gibraltar is certainly not an easy task.

Compact flock of Black Kites. These birds, like both stork species, Cranes, Griffon Vultures and Honey Buzzards, use common flying strategies on migratory journeys.

There are around 100,000 Honey Buzzards crossing via Gibraltar, compared to 25,000 via the Bosphorus and 39,000 via Messina. But we have absolutely no idea of the number crossing the Mediterranean either via the Balearics or across the open sea where they can't be detected. So the total migrating population must be really enormous.

The Black Kite

The Black Kite (*Milvus migrans*) is one of the most commonly occurring raptors on the planet. The largest populations in the Palaearctic are found in Russia, France, Germany and Spain. The European population is estimated at between 72,000 and 98,000 breeding pairs (Birdlife International/EBCC, 2000). In Spain it is found in the Balearics and on the mainland, especially in the western part of the peninsula, but not particularly in Galicia, the central area and down the Mediterranean coast from Gerona to the Strait. The current population is estimated at about 9,000 breeding pairs, although in 1970 Jesús Garzón put the figure at 25,000; the species, then, would appear to be in decline. They are trans-Saharan migrants wintering in tropical West Africa in Senegal, Gambia, Ghana, southern Mauritania, Mali, Nigeria etc. They can be described as sociable birds, although away from their normal breeding areas they may seem solitary. It is a common sight to see them gather at rubbish dumps and they breed in colonies, preferably in pinewoods where they fly and feed together in groups. On migratory passage they generally fly in large flocks, depending on the migrating season, and roosting sites usually contain several hundred individuals. Together with the White Stork, Black Stork and Honey Buzzard they are the migrant bird that flies in the most compact flocks. At their wintering sites in Africa they also get together in large groups to feed and sleep. In Mauritania "you get every single Black Kite from Spain", José Ramón Benítez said jokingly when he was monitoring the satellite transmitters of Egyptian Vultures migrating to Mauritania in 2003.

In Tarifa they used to call Black Kites the *wind birds* because the strong *levante* wind holds them up much longer than other migrants and their presence in the town is more conspicuous, but this description probably fits other same- or similar-sized raptors like Honey Buzzards and maybe Booted Eagles. Nowadays people call them eagles or little eagles. The English name comes from their tremendous ability to keep themselves in the air like a kite. When they are near enough to be observed with binoculars or, better still, a telescope, on moderate windy days you can appreciate their absolute aerial domination and by just moving their tail feathers they can maintain an almost static position. Their tail flips one way or the other and they stay parallel

Young Black Kite soaring on its way to a roosting site near Tarifa, where hundreds of them gather on very windy days.

to the ground without having to hover with their wings like other birds.

It is on prenuptial migration, if you find a suitable spot, when you can observe these skilful birds approaching the coast using their full potential in the face of very adverse crosswinds. What a sight it is to see these groups of kites flapping their wings non-stop, without taking a rest – just like the powerful Honey Buzzard – skimming over the foam-crested waves and only reducing the rhythm when they are close to the coast, which is when they alternate with soaring movements, possibly using the air currents caused by the swell, like seabirds do. Sometimes we've seen them gliding just centimetres above the waves, when, for every metre they go forward they drift five metres sideways, pushed by the crosswinds. People seeing this for the first time will have their heart in their mouth, thinking that at any moment the birds will fall, inexorably, into the sea. But this never happens, at least in all the years we have been watching prenuptial migration: despite fighting against all kinds of wind force, we've never seen even one fall into the sea. On the other hand, Short-toed Eagles and Griffon Vultures do fall in, and not because of harassment from gulls or other birds.

As with all migrants in the Strait, and depending on the wind strength and direction, flocks drift and come in via the Mediterranean or Atlantic sector. With westerlies they arrive at Tolmo-Faro, Gibraltar and even as far as Estepona, and, with easterlies, at Alelíes if they are light, or otherwise as far as Bolonia or Barbate. When winds blow from N, NW or NE they affect the central part of the Mediterranean sector and the Black Kite flies over at a considerable altitude, as do most other migrants under these wind conditions.

Black Kites resting on the ground after struggling all day against levante winds and being buffeted back and forth by strong gusts.

Black Kites lose their style and composure fighting against the almost permanent severe winds in the Strait and making progress, not to mention crossing, poses a real problem for them. But these birds are much stronger than most visiting ornithologists give them credit for. Sometimes you see them hesitating and struggling to maintain their positions near the coast and then passed by a flock of Honey Buzzards, yet they don't give up in desperation and suddenly, for some reason only the birds themselves know, they start a successful crossing.

The first Black Kites arrive early in the second half of February, although you may see a few stray individuals in the first half. And, as with all migrations of soaring birds in the Strait, one day you see one, two or three days later you see four or five, the number goes up to 15 or 20, but, suddenly, in come flocks of 100 or 300.

The flocks may start gathering on the Moroccan coast, varying from a few dozen to hundreds of individuals and they arrive at the European side in waves, like the White Storks or Honey Buzzards, depending on the wind and the time. On days when flocks come streaming over non-stop they start gaining height at the coast, regroup and then head northwards again at a considerable altitude.

On postnuptial migration Black Kites are masters at reversals, so that, when there are fierce *levante* winds, they come back to Los Lances, where they regroup and have another go at crossing. They are also very skilful floaters and you can see them doing this all day soaring up and down. On many occasions the wind wears them out and they land on some hill or other – they have their favourites – and spend one or more days there waiting for the wind to abate and then start their journey again.

As they are not good at flapping their wings to make rapid progress, they usually cross at first light when there is a *poniente* wind. The reader will recall that a characteristic of this wind is that it slackens off considerably in the evening and doesn't gather strength again until well into the following day; so the early morning is the time that Black Kites choose to cross the Strait with least risk to their well-being. This same tactic is the one we described for Honey Buzzards.

Nearly all migrant birds entering Spain come down the valleys of the western end of the Pyrenees and then some go along the Douro valley down towards Extremadura and others along the Ebro valley (De Juana *et al.*, 1988). After that the former will take a S or SSE course and the latter a SW or SSW. But as they approach the Campo de Gibraltar their flight course becomes dependent on the prevailing wind strength and direction in the area and they may be subjected to considerable drift.

Route taken by Black Kites across the Strait with a levante wind :

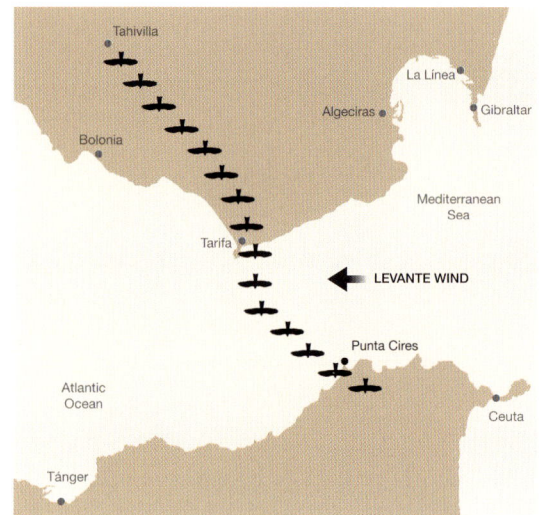

The approach to the Strait is very similar to that of the Honey Buzzard and nearly all raptors in general. With a strong, even moderate *poniente* wind they fly along the Málaga coast to the Rock of Gibraltar or cross the Bay of Gibraltar as far as the Punta Carnero lighthouse or just follow the coastline. If this wind is light then they approach from the northwest, go round the eastern edge of the *sierra* Algarrobo and turn towards the coast. It is the best species for studying the whole range of strategies used by soaring birds when crossing the Strait. In stormy weather they will keep floating in the air for hours, days even, unmoved by the adverse conditions and reversals are a systematic part of their migratory behaviour.

In 1972 I spent a day in the company of the prestigious French ornithologist J.M. Thiollay, who, after his experiences in the Strait, published *La migration d' automne à Gibraltar: analyse et interprétation* (1975) and *Importance des populations de rapaces migrateurs en Mediterranée occidentale* (1997). On that day we counted more Black Kites returning than those setting off on the crossing. The figures didn't add up because at the time we knew nothing about reversals and what we saw were birds on passage and others returning, further east, to try again. The truth is that this completely bamboozled us.

Approaches to the Strait with a *levante* wind blowing are made flying along the Cádiz coastline or a little inland, as we've seen before, depending on the wind intensity. If the wind is light they may come via the *sierra* Algarrobo, as we saw with the *poniente* wind. But with a fierce *levante* blowing in their faces you can see them toing and froing, going up and down in large, stretched out flocks, a really impressive sight when observed from the hill at Cazalla. Sometimes what ornithologists call a *wheel* is formed – that's when groups of birds are circling – and the force of the wind drags them inland until they regain control of the situation and start flying into the wind again.

The Black Kites' favourite crossing point, with both *levante* and *poniente* prevailing winds, is the Tarifa isthmus. From the city you can see them struggling tenaciously against the wind next to Palomas Island. And there you have another difference between Black Kites and Honey Buzzards: their preferred crossing points. Honey Buzzards prefer Tolmo-Faro, whereas Black Kites prefer Alelíes, especially Palomas Island. These patterns are, of course, upset by storms and the crossing can come to a standstill for several days. Even the actual time they leave depends on the prevailing wind. From reports by Bernis and from my own observations we deduce that Black Kites take advantage of any lessening of the wind's strength and then cross at the best possible moment, either early, in the middle or late in the day. At dawn with a considerably weaker *poniente*, Honey Buzzards and Black Kites are masters at grabbing their opportunity and crossing the Strait before the wind gets up again.

You can often see Black Kites flying together with other raptors or White Storks. Honey Buzzards don't do this, but that doesn't mean that there is a common strategy among different species to cross, only that they coincide in time and space. Finally we must point out that we believe that the first migrants to arrive in the area are the juveniles, judging by their plumage, which looks somewhat lighter and in more perfect condition in flight, although we have never taken the time to study this seriously.

The Red Kite

The Red Kite (*Milvus milvus*), is a scarce winter visitor but very rarely crosses via Gibraltar. We have never found a single nest of this beautiful bird of prey in the Campo de Gibraltar or La Janda. It is in October and November when they are to be seen in and around La Janda, before crossing the Strait and staying in Morocco. We have seen some, but very few, in September. On our visits to observatories in the area we have heard some observers – obviously those with little experience- claim to have seen a Red Kite, but they will have mixed it up with a Black Kite, most likely juveniles with prominent, light plumage. So we are somewhat sceptical when we hear a migrant Red Kite has been seen in the area in July and August.

Their approach and passage strategy is similar to that of the Black Kite, but the great difference is that it is uncommon to see more than two individuals together. I have only seen a pair arriving once or twice.

The Egyptian Vulture

The Egyptian Vulture (*Neophron percnopterus*) is found around the Mediterranean, the Middle East and Central Asia as far as India and, in Africa, in the southern Sahara and dry areas in the east and south (Donázar, J., 2003). The Palaearctic population, estimated at 2,900 to 7,200 breeding pairs, winters in Sahel Africa. In Spain it is found throughout the peninsula, in the Balearics and in the Canary Islands, where a new sub species has been recorded (*N. p. majorensis*; Donázar *et al.*, 2002). In peninsular Spain it breeds from the Ebro valley north- and northeastwards, from the Cantabrian Mountains to the Pyrenees; also in Extremadura, Salamanca, the north of Andalusia and in both the Natural Parks of Grazalema and Los Alcornocales. It basically feeds on carrion, but also on small animals. The population in the peninsula is estimated at 1,320 – 1,480 breeding pairs and all data we have agree that, over the last hundred years, it is a species in decline and in some areas, such as Andalusia, it is even scarcer than the Spanish Imperial Eagle. In Los Alcornocales, and more specifically on the La Almoraima estate, we knew of six pairs at the end of the seventies, but now there are none left and that is almost certainly due to the use of poison over the years on all game reserves. They don't acquire their adult feathers until their fifth year and breed only after they are four years old.

All Egyptian Vultures from Iberia, from just over the Pyrenees in France and the few pairs that breed further north, cross via Gibraltar. The few that breed in Italy cross via the Messina Strait

It is quite unusual to see a Red Kite migrating via the Strait of Gibraltar.

With its powerful wing beats the Egyptian Vulture is capable of crossing the Strait on very windy days when other species don't even attempt it.

Egyptian Vulture, one of the birds which suffer most from poisoning in the countryside, now struggling for survival in Spain.

and those that cross the Bosphorus these days have seen their numbers go down considerably.

On prenuptial migration we have never seen large flocks of Egyptian Vultures coming over the Strait. They generally travel alone and to see, say, a dozen individuals together, is quite exceptional. Like all migrants, they are affected by drift, although less so than most because they are excellent soaring birds and beat their wings strongly. On strong windy days it is not unusual to see them arrive at the coast, constantly flapping, and then spiral upwards to disappear from the watcher's view. Only on one occasion, as we mentioned earlier, have we seen an Egyptian Vulture on one of the cliffs, near Palomas Island, probably resting after becoming overtired. Most of the crossing, and certainly the last few kilometres, is with a slow flapping motion and hardly any gliding, and giving an impression of power and strength. The first arrivals are in the second half of February and this may continue until early May. I have seen adults and juveniles of all ages arriving, from first year up to sub adults, but I don't know in what proportion. They may come with other species, but this is just a coincidence in time and space, not a migratory strategy.

Where you do see relatively large flocks for this species, that is 20 to 25 individuals together, is on postnuptial migration. But rather than class them as flocks we see them as gatherings of individuals, especially when approaching the Strait, where they are sometimes held up by a fierce *levante* and float around for a time. After the Honey Buzzard it is perhaps the strongest flying bird and, if you will allow us to use the expression, the one which shows most determination to cross when conditions are adverse.

Approaches to the Strait are dependent on the prevailing wind strength and direction at the time. We have seen that many migrants show a preference on approach for the Atlantic or Mediterranean sector: White Storks go for the Atlantic sector and Honey Buzzards for the Mediterranean because they mostly come from northern Europe, but the Egyptian Vulture doesn't seem to show any particular preference for one or the other. This could be due to the fact that nearly all are Iberian vultures flying south from the breeding areas in the north, north west and north east and, as they are not affected by drift as much as other soaring birds, they fly in a straight line down to the Strait.

Unlike Honey Buzzards and White Storks they don't set off across the Strait from a particular spot. They fly along the coast and then head out across the sea, putting up with the buffeting from the crosswinds and flapping powerfully. Thanks to a test with new satellite tracking methods we know that out of six monitored birds, three did not cross via Gibraltar: two immature ones crossed from Estepona and another just east of the Rock. This proves, once again, that 100% of migrants do not cross via Gibraltar, the proportion of those that do varies – a possible exception may be the Griffon Vulture. As for its flying timetable, it doesn't set off very early, but prefers to wait for thermals to form and so passage is at its peak between 8 and 10 in the morning.

In 2002 an interesting study was conducted following an agreement between the Biological station at Doñana (EBD) and the Environmental Department of the Regional Government of Andalusia: *Conservation and Recovery of the Egyptian Vulture (Neophron percnopterus) in Andalusia: identification of wintering areas in Africa by satellite tracking.* Radio transmitters were attached to four chicks born in the province of Cádiz and two immature birds captured in nets in the Alcudia valley in the province of Ciudad Real. The six individuals migrated practically in a straight line, although three didn't arrive at their destination because they died on the way (two in central Algeria and a third in central Mali, not far from where the survivors wintered). The three that completed the journey were two immature birds – two and three years old respectively – and a one-year-old juvenile and they wintered in southern Mauritania on the Mali border.

Griffon Vultures

The Griffon Vulture (*Gyps fulvus*) is found in the Palaearctic and northwest Africa as far as central Asia, Arabia and Iran. The subspecies *Gyps fulvus fulvus* breeds in the Spanish peninsula and, after tragic and constant falls in the population until the 1960's, there was an increase of 80 – 90% between 1979 and 1989 in the same area of distribution (Donázar, 1987; Arroyo *et al.*, 1990) and a spectacular 506% between 1979 and 1999 (Martí, 2003). It has a special preference for limestone rocks and, except in the extreme northeast and northwest and in parts of Levante, Albacete and Huelva, it is found in the rest of Spain in varying numbers. The most populous vulture province is Navarra, followed by Cádiz and the largest colony nearest the Strait is near Tarifa at a place called Laja de Aciscar, where there are nearly 90 breeding pairs.

In 1772 the numismatist Francis Carter spent nearly two years in Gibraltar and travelled in the Málaga area. In his book – mentioned earlier and from which we took an extract to accompany the title of this chapter – he referred to the pre- and postnuptial migrations of Griffon Vultures in the Strait. The amazing thing is that there was no further reference to them until two centuries later when in 1969 Lathbury mentioned the passage of a total of 120 Griffon Vultures in the spring of 1968. One might be inclined to think that a numismatist's comments on birds are unreliable but if we continue reading his book it says and we quote: "The vultures that come from Africa every spring and pass over the Rock without stopping, return in autumn; they make their annual migration in flocks and, in flight, are easily distinguished from storks (which are also birds of passage) as they carry their feet under their tail, whereas the storks' feet hang downwards. There is a vulture in the garrison that, I suppose, stayed on the Rock out of tiredness; it is large with beautiful plumage; its outspread wings measure eight feet; its rump is wide, tall and covered in soft, shiny, dark feathers; these birds

The Griffon Vulture, together with the White Stork, is one of the migrant birds whose numbers crossing the Strait have increased spectacularly over the last few years.

Group of adult Griffon Vultures roosting in the Valle del Santuario (Tarifa).

can go without meat for a long time; they prefer carrion, which they eat voraciously, to fresh meat." My personal opinion is that it would be an immature bird because an adult has a white throat and he would surely have mentioned this as it is one of the most striking characteristics of the Griffon Vulture.

While I was busy with my hobby, scuba diving, from the end of the 1950's until the mid seventies, my contacts with bird migration led me to see vultures arriving at the coast of the Strait and I remember seeing the body of one near Calarena, next to the Punta Carnero lighthouse,

but it never occurred to me that these enormous birds were actually crossing the Strait on two migrations. So, when I found out, I was really surprised because I had witnessed something that most experienced ornithologists weren't aware of, and I just thought they were carrion-eating birds flying over the beaches searching for large, dead fish, dolphins etc brought in by the tide.

Before continuing with this subject, the migration of Griffon Vultures via Gibraltar, let's examine other types of behaviour in the species, which may be a source of error or misinterpretation, especially for those observers who live further inland.

These birds take part in nomadism and other wandering movements (Bernis, 1980; Del Junco & Barcell, 1997) which we are going to deal with to enable migratory movements in the Strait to be better understood. Bernis says, and

I quote, "*nomadism involves a movement result-ing in a prolonged stay in a new area of at least a few weeks, if not months, and this movement isn't necessarily a regular one in time and space, as it can be in different directions, and this is what differen-tiates nomadism from real migration*". As for the wandering movements, these are classed as "*rou-tine or occasional movements within the same day or over several successive days and nights, returning to the habitual nesting or dwelling area.*" Some of this nomadism is related to transhumance and the wandering movements to the daily search for food.

Just like the White Storks, the Griffon Vul-tures of western Europe - the Iberian popula-tion and some French ones from the north side of the Pyrenees and the Massif central – cross via Gibraltar and winter in Senegal, Mali and Niger, whereas those from eastern Europe mi-grate via the Bosphorus and the Suez Canal to winter in Sudan and Ethiopia.

In over 30 years of watching Griffon Vul-tures reach the coast in prenuptial migration I can categorically state that the numbers cur-rently arriving are really spectacular. In the 60's I can't recall seeing any sizeable flock, although it's true that I didn't make a note of them and my interest at the time wasn't strong enough to make me get my binoculars and go to the coast solely to watch them coming, because I knew nothing of their phenology. In the 70's and 80's you could spot flocks of 40 or more and in the 90's there was an increase both in the number of birds and flocks in the migrating seasons. Nowa-days it's easy to see flocks made up of 90-100 individuals and I've even seen one of 286 and another of 317. In fact, in just a single day you quite regularly can see more than 200 vultures spread between two or three flocks.

Rarely does a Griffon Vulture cross the Strait on its own and whenever I've seen one coming over alone I've had a good look through the binoculars and usually spotted the flock fol-lowing behind or drifting because of the wind. There are no figures for these vultures coming on prenuptial passage because nobody has con-ducted a study from March to June, the months this migration lasts, but from sporadic data I have obtained and some from my friend Pablo Ortega, there may well be over 2,000.

They generally cross with prevailing *levantes* and *ponientes*, both from the West and North-west. I have never seen them arrive with a SW wind. Another peculiarity is that, as in postnup-tial migration, they choose to cross when winds are, at most, moderate. I haven't heard of vultures arriving in a *poniente* or *levante* of more than force 6 on the Beaufort scale and that may be the maximum they could tolerate with a drift taking them well beyond the Rock with a *poniente* and over to the Bolonia inlet with a *levante*. When there's a force 4 or 5 *poniente* they come in ex-hausted over the eastern part of Tolmo-Faro and much nearer Punta Carnero lighthouse, the Bay of Gibraltar and the Rock. These 20 kilometres or so of sea that they have to fly over, flapping continuously, partly because of drift, seem to be this species' physical limit. If the Strait were a few more kilometres wide it could be that not a single one would manage to reach the other shore alive in a moderate wind. Countless times I have seen vultures collapse onto the beach, ab-solutely spent and unable to keep their wings going because their exhausted muscles couldn't stand their weight.

In 2005, in the month of June to be precise, I saw at first hand and for the first time how exhaustion had caused the death of a flock of

Griffon Vultures. After several weeks of strong *levantes* 11 of these birds were washed up, drowned, on the beach between Valdevaqueros and Punta Paloma. They were all immature and were washed up on the same day. This makes us think that they fell into the sea near the coast having been affected by the *levante* drift, against which they must have had a titanic struggle. When they ran out of strength they fell to the sea and died of hypothermia. Possibly because of the nervousness and tension built up during weeks of waiting for favourable winds on the Moroccan side, one of them decided to chance its luck and dragged the rest with it.

The crossing is in flocks, but these aren't as compact as those of the Honey Buzzard or Stork, because on arrival, when we first see them, exhaustion has taken its toll and the weaker ones have been relegated to the back. The flock is somewhat dispersed compared to the solid unit that must have left the Moroccan coast. A compact formation is what you see when the postnuptial flocks set off for Morocco. At Tolmo-Faro when there's a 3 or 4 force wind, flocks of Griffon Vultures arrive in formation, and even in Indian file on many occasions. At the Punta Carnero lighthouse they often arrive from the SW or SSW and on isolated occasions from the West in a force 4 or 5. The explanation may lie in the drift that pushes them towards the Rock but when they're getting towards Punta Carnero and fail to reach the coast, then drift makes them fly another 10 km, which is the distance to the back of the Bay of Gibraltar, or another 7 km to the Rock of Gibraltar. This means that if they change their flight path from due north and head west then they will reach the Punta Carnero area with less effort. From that position you can see them arrive as if they were coming from the Rock.

Another reason to consider is what I call redirected gusts in the Bay of Gibraltar. Many times when I've arrived at Punta Carnero there's a *poniente* blowing, whereas in the Bay of Gibraltar it's a *levante*, as you can see from the direction of the smoke from the chimneys in the industrial complex there. The vultures drift towards the Rock when there's a westerly in the Strait but suddenly, when they're in the Bay of Gibraltar, they are faced with a *levante* which pushes them towards Punta Carnero or to the whole of the western side of the Bay, so then they decide to go for the lighthouse area or that stretch of coast between the lighthouse and Getares beach.

Sometimes, with a northwesterly blowing, Griffon Vultures circle upwards over the sea for a few kilometres before reaching terra firma and I think they let themselves be dragged by the wind up to a certain height before, heading into the wind with outspread wings, they decide to float down to the ground. This is the best wind to cross with because they fly into it and soar practically all the way, or at least most of the time and alternate soaring with short flapping intervals. They arrive at the coast flying very high and apparently not very tired. However, when there are moderate *poniente* winds, force 4 or 5, you see them arriving worn out, bills open, like athletes gasping for breath.

I can assure readers that the prenuptial migration of Griffon Vultures over the Strait of Gibraltar is without doubt one of the high points of migration in the area. From far away you can make them out by their slow, rhythmic flapping and the generally medium altitude of their flight path, except on days when there's a prevailing north wind. When we are standing at a height which gives us a first rate view of them coming towards the coast, it's an unforgettable moment, and not just for an ornithologist, as they pass close by, at the end of their strength, bills open, but then gaining great height in a short space and rising above us. There are few migratory sights, at least as far as I am concerned, that cause such admiration, if you are fortunate to know the ideal place where you can be so close to them.

As they are not so adventurous as to cross with strong prevailing winds – the same happens on postnuptial passage – drift doesn't affect them as much as other raptors that, like Honey Buzzards, rely on their strength and cross with strong winds that force them to drift several kilometres in one direction or another. Wherever there are steep cliffs or hills that fall abruptly to the sea, slope currents form and help the most exhausted birds on towards the hinterland. They are a life-saving element for vultures, which can hardly spread their wings to stay in the air, and

quickly lift them up to greater heights, enabling them to continue their journey to unknown destinations.

From the Rock, Lathbury recorded adults arriving in spring but after looking at hundreds of photos of vultures arriving, some of which are of just the breast and one wing, I haven't found one showing a white throat ruff which would absolutely prove it is an adult. We know that that doesn't determine if the individual is a breeding bird or not, because I do have photos of breeding birds with feathers instead of down on the throat ruff. With binoculars it's not easy to distinguish them at the distance the observer may be from them in the open countryside but, at certain points in the Strait, vultures are just a stone's throw away and you would clearly be able to see the white throat ruff of some of them. I deduce from this that those arriving on prenuptial migration can only be categorized as subadults at most, or with the external appearance of this category. Another fact is that arriving at that time, even if they were adults, they wouldn't be able to breed and would have to wait nearly a year to be mature enough to mate. The odd adult or two may arrive on this migration but it is not usual.

There must be fewer birds returning to the Peninsula than leaving as there are a considerable number of deaths, but in spite of that the species is enjoying an upswing, so logically every year there is an increase in numbers leaving Iberia to winter in sub-Saharan Africa. It is there that many die on their journey or at their destination because of inexperience, but nevertheless more are returning now as the breeding population rises. The juveniles spend five years in Africa before returning, which is when they are ready to breed. Unquestionably, those Grif-

Routes followed by Griffon Vultures across the Strait :

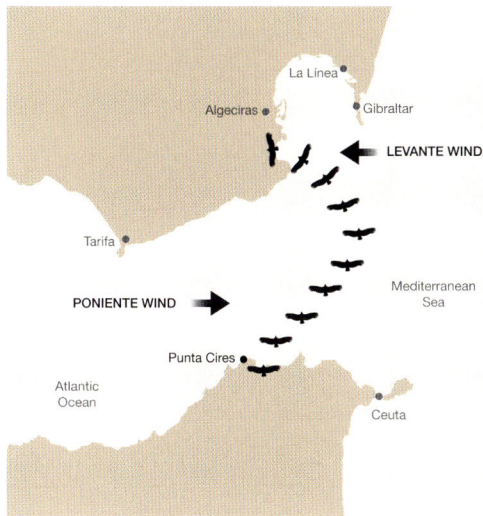

Griffon Vultures feeding at the Campo de Gibraltar refuse dump on a Sunday, when there are few vehicles about and they feel more confident.

fon Vultures returning on prenuptial passage are not from Morocco as there is no presence there. On several expeditions to the Atlas Mountains and the foothills no sightings of these birds were recorded, although Ospreys were. I know many ornithologists and not one has ever seen a Griffon Vulture in Morocco, so we can conclude that those that arrive in spring are from the Sahel.

According to Bernis (1980) the crossing of the Strait is irregular and spread out in time. But for me the most surprising data, when compared with up-to-date records, are those provided by Pineau & Giraud-Audin when they were monitoring the Moroccan side from February to June, both inclusive, and only detected 12 passage days, the largest flock being made up of 60 individuals. As we have already mentioned, 60 migrants in a flock is now quite commonplace and we have seen one of 286 and another of 317 in 2003.

In 1971 the first studies of vultures crossing during postnuptial migration were conducted by Pineau & Giraud-Audin, and, together with 1973 and 1974 yielded a grand total of 599 birds crossing. Thiollay and Perthuis also studied the passage of vultures in 1974 and counted 422. The studies led by Bernis in the 70's ended in October, yet it is precisely in early November when passage across the Strait is at its peak, so the figure of 734 birds crossing in the autumn of 1976 is therefore incomplete. It must also be pointed out that those were the years when the vulture population started to increase, after decades of agonising decline of the species. Above we gave the numbers of Griffon Vultures crossing the Strait as 2,000, according to studies conducted by Joaquin Griesinger, whose study exercise at the beginning of the 90's was incomplete. At the moment the Migres Programme includes this bird in its studies and has reported 1,386 crossing, a figure we feel falls short if judged by those Griesinger counted in the mid 90's, but it could support the theory that some young vultures stay in the area (as nomads) and feed at the Los Barrios rubbish dump in autumn and winter. But this doesn't resolve the, in our view, paradoxical situation in that more birds return in spring than were counted out in autumn.

In the Strait you can see – and it's an undisputable fact – vultures on postnuptial migration and also a lot of vultures toing and froing from the nearest hinterland; but we now know that nomadism is common in the area, i.e. there are vultures that, instead of migrating across the Strait, winter in the area. They are classed as wintering birds, as are some White and Black Storks in many areas in Iberia. So, not all Griffon Vultures seen in the area during autumn migration are really migrants waiting to cross the Strait. In a study carried out by J.R.Garrido and Cristina G. Sarasa at the Los Barrios dump in 1997 they monitored 22 ringed Griffon Vultures whose origin was as follows: 4 ringed in Andalusia at recovery centres near the Strait; 3 from North Spain and 2 from the South of France. The origin of the other 13 birds was not clear from their rings. Of all these ringed birds 40% didn't cross the Strait and, of these, 2 were observed in successive years. Between 40 and 60% of the birds studied were not adults.

Just as we have already observed that Honey Buzzards, Storks and Black Kites have their favourite routes to approach the Strait, so we must wonder what vultures do. This is not an easy question to answer because you can see vultures flying in the area constantly all the year round. Even if a flock is approaching, unless it is unusually large of course, an observer might think it is a group exploring the territory. Over every town and even large city in the Campo de Gibraltar there is a constant toing and froing of vultures in the sky. What I mean by this is that it is difficult to detect whether flocks are arriving in the area or if they do it like this. Possibly they start flying south on their own and gather into different sized groups along the way.

The *apathy* with which the Griffon Vulture faces up to the crossing isn't found in any other soaring bird in the Strait. They can wander up and down the coast for days, even those days the observer feels are ideal for the Strait crossing, with light *poniente* or *levante* and northwesterly winds. In spite of this the flocks, or one large flock, carry on circling, gaining height and floating parallel to the coast as if they were in no hurry to leave. Those of us who have followed these movements closely get very frustrated because it looks as if they're never going to cross. Sometimes four or five days of favourable winds go by and, although you feel they're going to cross at any minute, they don't, but when the ornithologist least expects it, they head for the Strait, leave the coast behind and fly determinedly to the Moroccan side. No Griffon Vultures have been seen crossing on postnuptial migration with even only light *levante* winds, nor have they been spotted with *poniente* winds of more than force 6.

Pineau and Giraud-Audin's studies in postnuptial migration in 1975 at the Moroccon side of the Strait only recorded vultures arriving on five days between 23rd October and 19th November. This behaviour, of flocks crossing in spurts in autumn and then blank days despite ideal conditions, doesn't occur in prenuptial migration. At the end of April and, mainly, in May on the best days for it (a *poniente* maximum force

6, or a northwesterly) flocks of Griffon Vultures cross practically every day. Nowadays I think the dump at Los Barrios has some influence in autumn, because the vultures find food there without having to search for it. And possibly this dump is so attractive that they decide to winter there, as we pointed out earlier (Garrido & G. Sarasa 1997).

As the reader can see, figures are larger or smaller, depending on who gives them, and although they've been gathered over more than 30 years, they may be influenced by the areas chosen to do the counting, both in spring and in autumn. This may be the key to the different interpretations and we will deal with that in Chapter VIII (The Ornithologists's Guide to the Strait).

What most authors do agree on is that the crossing is via the Tolmo-Faro sector, at the Tolmo inlet. It really is an impressive sight to watch a flock of, say, 81 Griffon Vultures on one of its coastal "wanderings" circle upwards in a thermal and, suddenly, off go a few towards the sea to be followed by the rest of the flock, all flapping their wings, which, until then, they had only used for soaring. A sight like this is unforgettable for an ornithologist, as he sees these enormous birds moving together with powerful, synchronized wing beats. On one particular day in the mid-eighties members of the former GEODE organized a trip to the coast to count waders and we were all captivated by such a magical moment.

However, in the Strait no rules are followed 100% and the postnuptial migration of vultures is no exception. Joaquín Mazó and Guillermo Doval , from the Migres Programme, related to me how in September 2005 a flock of about 20 birds crossed towards Morocco from Alelíes, something quite unusual for the time and place. They described how a vulture from this flock followed a Short-toed Eagle that started crossing and the rest of the flock were "dragged" across after their companion. This makes us think that, at the end of September and in early October, when the migration of raptors is ending and Griffon Vultures are just starting on theirs, some flocks allow themselves to be "dragged" across the Strait. Once these "hares" – to use an athletics term – have disappeared, then the vultures resume their typical reluctance to cross. Possibly, and perhaps in the month of August too, this is what happens to local young vultures which are beginning to disperse along the European side of the Strait.

How far do the French and Iberian vultures get on their African journey? The data we have relates to the 1979 – 1987 period. Four chicks were ringed in the province of Cádiz. The two birds ringed by ornithologist Javier Alonso near Alcalá de los Gazules were recovered in Senegal. The other two were ringed by members of GEODE, one in Algar, recovered 184 days later in Morocco and the other in Los Barrios, again recovered in Senegal. It appears from the data we have that Griffon Vultures migrate along the Atlantic coast of Africa so as to avoid the tremendous heights of the Atlas mountain range, which are east-west oriented, and, in doing so, they make good use of the thermals which form near the sea. (Del Junco & Barcell, 1997).

The Short-toed Eagle

The Short-toed Eagle (*Circaetus gallicus*) is found in the Palaearctic from Iberia to India, nesting in South and Central Europe, the Caucasus, the Middle East and in Central and Southern Asia. In Africa it only breeds in a narrow strip to the north of the Atlas Mountains (Mañosa, 2003). The estimated European population is between 6,200 and 14,000 breeding pairs, the majority of which are Spanish (according to Birdlife International/EBCC, 2000) and the Iberian population is between 1,700 and 2,100 pairs (De Juana, 1989). The last census produced an estimated 2,000 – 3,000 breeding pairs (Mañosa, 2003). As we can see, these figures differ quite considerably. The species is found in most parts of Iberia, but to a lesser extent in the North West in Galicia and in the north of Portugal. It is not present in the Balearics or the Canaries. In the area near the Strait it is, along with the Booted Eagle, the most abundant forest raptor.

This forest species needs wide plains near its nesting site where it can capture the reptiles it feeds on. Nests are often built on the edge of the forest and in the Campo de Gibraltar nearly all are to be found in smallish cork oak trees on sunny, north-facing slopes.

Together with the Griffon Vulture it is the most outstanding of all soaring birds, but that is not its only quality: unusually for its large size it can hover in the air when hunting. Another technique it uses is to stalk its prey from the top of electricity pylons, although the consequences for the bird itself can sometimes be negative!

Prenuptial migration begins in the second week of February, although in some years it has been delayed until the end of that month, and it is common to see them arriving with flocks of Black Kites, crossing the Strait at the same time. They don't form flocks as such, rather we should speak of more or less dispersed groups or collections of birds which join up as they cross. Short-toed Eagles travel alone, but as they approach the north of Morocco they coincide with others of the same species. When thermals warm

Opposite

Head of a Short-toed Eagle. Its sharp beak and exceptional eyesight are essential assets to be able to follow a diet based almost exclusively on reptiles.

A young Short-toed Eagle fighting desperately against a large greenish-yellow Horseshoe Whip Snake *(Coluber hippocrepis)* which bites one of the bird's wings.

up the ground there, all the Short-toed Eagles that have roosted in the area start to use them and begin crossing the Strait if conditions are favourable. This is why observers on the European side see them arrive in dispersed groups, not forming a flock. The physical state of each bird influences the extent of this dispersion; probably the weakest ones are the stragglers. Just like Black Kites they are masterful soaring birds, but in strong winds you see them suffering to make progress.

A group of Short-toed Eagles crossing the Strait does have something in common, but there is not nearly as strong a link between them as in other soaring birds, such as storks, vultures, black kites etc. The most advantageous winds for them are westerlies and north westerlies. A westerly of force 5 will make them drift a lot towards the Rock and with a force 6 not many will attempt the crossing. But, if there's a north westerly, then conditions are perfect and they will fly into the wind and show their skill as great soaring birds. When this wind is blowing they set off from the Moroccan coast near Tangier at a great height and hardly lose it as they cross the Strait. They pass over the European side at altitude, hardly flapping at all and this is quite different from what they do when there is a *poniente* wind. It is second only to the Griffon Vulture in the difficulties it experiences in crossing the Strait, although if there were no crosswinds this adventure would be no trouble at all.

It is important for the reader to understand the great dangers that migrants have to overcome when choosing to cross via the Strait of Gibraltar because the difficulty doesn't lie in the 18 km or so which separate the two shores but in the strength of the winds which blow every day and, to make things worse, in the fact that these winds blow sideways on to the birds' course. Consequently Short-toed Eagles don't cross when there's a strong *levante* and with moderate winds drift takes them well off the intended route. Many a time one sees them arriving with their bills open and wing beats that are not as wide and spaced out as they are normally, a sure sign of stress and tiredness.

From the shores of the Strait you see groups of them arriving with their characteristically slow, heavy flapping. From a distance, when there is heavy sea mist they can be mistaken for Griffon Vultures, but with a bit of experience you soon learn to identify them even when visibility is very poor. Just as it is unusual to see a Griffon Vulture crossing the Strait alone, maybe because they are gregarious birds by nature, in the case of the Short-toed Eagle it is not at all so. Very often, especially in the month of May, one will arrive on its own and not be followed by another until much later.

My personal record of Short-toed Eagle sightings – in three and a half hours with a force 3 to 4 wind – is 524 on 11th March 1991, but there were certainly a lot more. On that day I wrote down in my notebook 238 until 11.30 a.m., of which 101 were flying together in a thermal like a corkscrew or *rosca* - the Spanish term used to describe a group of birds within a single thermal -, although the casual observer might have thought they had arrived in a compact flock.

All nature photographers will agree with me that you can't take photographs and do another job well at the same time, because photography needs all your attention. You have to concentrate totally on the job you're doing to get a good result because, apart from focussing on the animals, in this case the Short-toed Eagle, you must choose the speed, the f-stop, the light

value, the sensitivity (both for slides and for digital support) and many more details that we will look at in the final chapter. If we add to this that you have to guess or, rather, sense what direction a bird is going to take an instant later, then it's a difficult and stressful task. In a nutshell, photography is not compatible with counting birds

Short-toed Eagle flexing its tail to balance its body on a very windy day.

After arrriving late and spending the night near the Punta Carnero lighthouse, a migrating Short-toed Eagle sets off again.

passing through your field of vision, whether it be overhead or to the left or right. I'm absolutely sure that thousands of birds could have crossed, because when I had to leave they were still passing and there were still several hours of daylight left. You never forget a day like that.

I have seen these splendid eagles arriving and collapse onto beaches, totally spent after the rigours of the crossing, or fall into the sea or be attacked by other birds… But I've never seen one arrive with a snake's tail sticking out of its bill. But my friend Pablo Ortega has, and he told me of a time when a Short-toed Eagle arrived on these shores and, from the air, dropped the snake it was carrying. Such an unusual incident makes me think of pilots who have to alleviate the weight of their plane to save fuel and be able to make it back to base or balloonists who have to do the same to avoid falling into the sea. The simile is reasonably accurate.

Black Kites and Honey Buzzards arrive at the shore of the Natural Park of the Strait at dawn, but Short-toed Eagles must wait until thermals have formed in order to gain height and, after a time which varies depending on the prevailing wind, decide to cross. As a result it's very unusual to see one arriving before 8 a.m. and arrival time is generally between 8 a.m. and 1 p.m.

It is quite normal for studies conducted in the Strait to concentrate on postnuptial migra-

This Short-toed Eagle stares at the photographer as it glides smoothly along, quite near the Guadalmesí bird observatory.

tion and the Short-toed Eagle is no exception. So here we will describe our modest experiences along with those of others, such as Bernis, who has devoted years to the study of postnuptial migration, although his published work has still not been given the attention it deserves in the area.

As happens to nearly all soaring birds approaching the Strait, crosswinds make them drift, but they travel further inland and that is why they are hardly sighted on the Rock in autumn. If they arrive in the area in the evening they will spend the night in the woods nearest the Strait. Once we found a roosting site of several dozen birds in a cork oak forest, on trees that were quite far apart. Short-toed Eagles like to mark out their own territories and even on migration they are intolerant towards those of their own species. Anyway, it must have been a one-off because we have never seen a roosting site again anywhere in the area.

A westerly wind sees them arriving in the area along the Málaga coast and further inland than other species. But some do cross the Bay of Gibraltar then fly between the south east side of the *sierra* Algarrobo and the coast, although the general tendency is to prefer the interior. They change direction coming from the north east and veer south west to fly over the Mediterranean sector, crossing from Alelíes if the wind is favourable. If there is an easterly blowing, then they manoeuvre in a similar way but over the Atlantic sector and leave via Alelíes.

Their attitude when it comes to crossing is not the "let's not bother today" one of Griffon Vultures, nor do they have that decisiveness of Honey Buzzards. The Short-toed Eagle's behaviour is more like that of the Booted Eagle or even the Black Kite. When there is a strong *levante* they like to soar high into the wind and there are a lot of reversals. You can see them arriving from the top of Cazalla Hill flying towards Los Lances beach high above the observatory and hardly flapping at all and just when it seems they are going to cross they let the *levante* carry them back towards Los Lances and they start the cycle all over again. These reversals seem to go on for as many times as it suits them. Once when I was at Cazalla I saw one pass over six times – it had a feather missing from each wing, so was easy to identify.

We have commented that on prenuptial migration groups of Short-toed Eagles don't arrive in compact flocks, but rather in groups of individuals which coincide at the bottleneck in the Strait. On postnuptial migration, however, when we are able to study this from another perspective, we do observe that the flocks or groups of dispersed individuals hardly mix with other species like Booted Eagles or Black Kites with which they coincide (although there are fewer and fewer), so you could say that here there seems to be a certain, albeit loose, relationship between Short-toed Eagles.

The main crossing times are between 8 a.m. and 2 p.m., hardly at all before that time and only in small numbers afterwards. In the afternoons you see a lot giving up and flying back to the woods inland to spend the night.

On 29th July 1996 a satellite transmitter was attached to a juvenile Short-toed Eagle near the French town of Cognac. It started its migration on 25th September, arriving at Gibraltar seven days later on October 2nd. It then went via Morocco, Algeria and Mali, to winter in Niger. The Cognac-Strait leg of the journey was in an almost straight line until it reached Jaén from where it continued south west. It flew on to Mali, also on a straight course as far as the 15º N parallel where it turned eastwards, arriving in Niger flying in a south easterly direction.

Harriers

The Harriers we are going to deal with in this book are those of the Circus genus: the Hen Harrier (*Circus cyaneus*), the Marsh Harrier (*Circus aeruginosus*) and Montagu's Harrier (*Circus pygargus*). This group of birds doesn't usually migrate across the Strait of Gibraltar. As they are exceptionally gifted soaring birds they can cross the Mediterranean from any point they choose. Those seen in the Strait are only a small fraction of those that migrate and if you add to the small number of harriers detected the fact that they are difficult to distinguish, especially females and juveniles, then observation data have to be treated with some caution.

The Hen Harrier

The Hen Harrier (*Circus cyaneus*) is a Holarctic species found in Iberia from the Pyrenees right across to north Portugal and the northern part of the Iberian mountain range as well as in pockets of Madrid, Toledo, Ciudad Real and part of Extremadura. It nests on the ground, as does Montagu's Harrier, which constitutes a danger for the survival of the species, especially in cereal-growing areas, although as it is a late breeder this particular danger has less of an impact than it has on Montagu's Harrier. There are fewer than 800 breeding pairs in Iberia. In winter some harriers from outside Iberia spend time dispersed throughout the south. On prenuptial migration the first male birds are thought to arrive in March and April. According to Bernis it is not a trans-Saharan migrant and that explains the low quantity of individuals observed in the Strait, with a dozen or even fewer birds on postnuptial migration. In 1997 Migres Programme observers only recorded one and the following year just five Hen Harriers. For our part, we have never seen it on prenuptial migration and the largest number recorded was 86 in the Strait of Messina in 2001.

The Marsh Harrier (The Western Marsh Harrier)

This species of Marsh Harrier (*Circus aeruginosus*) has colonized both hemispheres. It is found all over Europe except in the coldest regions of Scandinavia and Russia. In Spain it is found both in the Balearics and on the mainland in the Douro, Ebro, Tagus river basins and in the Guadalquivir marshes, its population estimated at more than 800 breeding pairs. The Iberian population is sedentary as is part of the European population. It is a trans-Saharan migrant that winters south of the Sahara, crossing via Gibraltar in the months of September and October and even in November (when 29 were counted in 1997 and nearly 50 in 1998 according to the Migres Programme, whereas Bernis recorded 355 in 1972 and 119 in 1977). Prenuptial passage is in March and April, flying individually, not in flocks. In the Strait of Messina the largest number recorded was 3,069 in 2003. From these figures and its good navigational skills it can be assumed that it may cross the Mediterranean without necessarily using the straits. But, although its skill as a flyer is well known, it is common to see them hesitate and even not cross over to Morocco and then wander about for a whole day before deciding to cross. When it does cross on postnuptial migration it will not start from any special place and it will fly at any time of day. It usually flies alone and on rare occasions you may see two together. Pineau & Giraud-Audin have recorded seeing 45 harriers at a roosting site in Morocco but it is very unusual.

Montagu's Harrier

Of the three harrier species crossing via Gibraltar Montagu's Harrier (*Circus pygargus*) is by far the most numerous and has the largest population in the vicinity of the Strait. In the Palaearctic there are an estimated 30,000 to 45,000 breeding pairs and in Spain about 5,000. In Europe it is found especially in the Iberian Peninsula and France and there are a few in the Balearics too. It is an occasional visitor to Madeira and the Canary Islands. It is present in nearly the whole of Spain, although less towards the east and in the northern part of the Cantabrian Mountains. It winters below the Sahara from Senegal as far as Ethiopia and down the eastern side of the continent

Marsh Harrier exploring La Janda, on European soil now after its migratory journey.

to South Africa. Because it nests on the ground, usually where cereals are grown, it is very vulnerable to predators and agricultural machinery. To reduce the number of chicks falling foul of combine harvesters and to increase awareness of the causes of the decrease in population, advertising campaigns are being stepped up.

Like the other harrier species migration isn't channelled across the Straits so the birds crossing via Gibraltar are only a fraction of those crossing the Mediterranean as a whole. They reach the European coast in March and April, the first arrivals being the males and all tend to fly over within a few days. They arrive on a wide front and don't aim for a particular spot. On numerous occasions I have seen them rise up to a great height when the coast is only a short distance away and be dragged into the hinterland by the wind. Bernis brilliantly describes the migratory flight with strong prevailing winds thus: "You see it drifting past being pushed sideways just like a paper doll in the wind".

Postnuptial migration takes place between mid August and the first half of September. Passage data collected by the Migres Project in 1998 don't compare with those collected by Bernis in 1980. While Migres only recorded a few hundred birds crossing, Bernis counted on average 1,316 over four years. In our opinion, when observing harriers in general, they tend to travel alone, sometimes at high altitude and you need observers who are very well informed, alert and under supervision so as to be able to monitor the passage of these birds as closely as possible. Furthermore, according to Bernis, local birds are often mistaken for migrants by observers who have little experience in the area. In 2000 the number of birds counted crossing the Messina Strait was 866.

Male Montagu's Harrier, which doesn't need to cross the Mediterranean via the straits as its powerful wing beats enable it to travel very quickly.

The Sparrowhawk

The Sparrowhawk (*Accipiter nisus*) is a small, mainly forest-dwelling raptor and therefore linked to this type of ecosystem in its distribution throughout Eurasia. It is found in all European countries except Iceland. Birds from central Europe winter in Iberia between September and November, although those from the north are fewer in comparison. (De Juana *et al.*, 1988); Iberian Sparrowhawks are sedentary. The European population is an estimated 280,000 to 380,000 breeding pairs (Birdlife International/EBCC, 2000). In Spain it is absent from the Balearics, the last known bird disappearing from the island of Menorca in 1970, and in the Peninsula there are between 6,000 and 10,000 breeding pairs and 150-200 pairs in the Canary Islands (Balbás & González-Vélez, 2003). It is present practically all over mainland Spain, especially in the

In a cornfield a Montagu's Harrier chick, not yet fledged, waits expectantly for an adult to bring food.

A Sparrowhawk flies hurriedly past the photographer's lens on postnuptial migration which will take it into unknown territories.

north, but scarcer in a wide area which includes Extremadura, Huelva, Sevilla and Córdoba. It is a prolific breeding species and may succeed in rearing five chicks, a presage of the high death rate among juveniles.

Prenuptial migration begins in mid January and lasts until the end of May. Their arrival is quite discreet for several reasons: they are small birds, travel alone and fly very close to the ground when over land. Given that they are competent flyers they probably don't need to cross via straits and those that come via Gibraltar are only a small portion of the total number of migrants. You must be very watchful indeed to be able to spot this little raptor on prenuptial passage as it may be flying at a great altitude or, if observers are standing on a hill near the coast, it may pass at knee height!

When there are strong *poniente* or *levante* winds blowing you see them coming from Morocco low over the sea, beating their wings strongly and not drifting as much as other birds. They are not considered to be pure soaring birds because, although they do use thermals to gain height, they habitually fly close to the ground, alternating a series of continuous, rapid and vigorous flapping with gliding or soaring. Like the Honey Buzzard these small raptors are less afraid of strong winds and cross the Strait at times when other species hold back, so that drift may take them as far as Barbate or Estepona.

When the wind drops they usually fly much higher - especially with a northerly or, preferably, a northwesterly – and then they are just pinpoints in the sky. According to Bernis (1980) first year juveniles do migrate but this tendency gradually disappears as they get older.

On postnuptial migration detecting the Sparrowhawk doesn't get much easier because they approach the coastline flying low over the hills. At the same time detection is complicated further by the presence of local Sparrowhawks which hunt by flying very low over the Spanish gorse shrubs, the dominant vegetation in the Natural Park of the Strait. These local birds fly back and forth over their territory again and again, skimming the shrub tops and catching small birds unawares. This means that, if the observer hasn't been well informed, they might count the same Sparrowhawk several times, as if it were a migrant.

Wind direction doesn't seem to have much effect on them, unless the wind is really strong. They cross at any point in the Strait and even at places outside the provinces of Málaga and Cádiz. If, as we have mentioned, the Egyptian Vulture crosses as far away as Estepona, these birds are guided by instinct and don't seem to mind where they cross, having little or no fear of flying over the sea.

The Buzzard

The Common Buzzard (*Buteo buteo*) is a medium-sized raptor found in almost all the Palaearctic except in parts of Ireland and Iceland. It is sedentary in Europe. It is a migrant in Asia and northern Europe and partially in central Europe (Balbás, 2003) and is one of the most abundant European raptors with a breeding population of between 690,000 and 1,000,000 pairs (BirdLife International/EBCC, 2000). In Spain it is found in the peninsula and the Canaries but not in the Balearics, the population being 13,000 to 18,000 breeding pairs (Balbás, 2003). In the Canaries it is not found on Lazarote but on the other islands there is a sub-species *B.b. insularum* with 430-445 pairs. On mainland Spain it is only absent from a large part of Almeria. A large part of the Scandinavian and central European population winters in Iberia (De Juana *et al.*, 1988).

During prenuptial migration you could say its presence is rather discreet as it doesn't travel in flocks, but individually like the Sparrowhawk. The most I've seen in one day has been six, but spread out over three hours and I've only ever seen two together once. So my experience of prenuptial migration is very limited and there's not much I can contribute. On arrival they fly in a similar fashion to the Honey Buzzard but without such prolonged soaring stints prior to a series of powerful wing beats. When they are circling upwards Buzzards keep their wings at a sharp angle, not as flat as the Honey Buzzard.

A Buzzard, an increasingly rare bird of prey around Gibraltar, possibly due to climate change.

Postnuptial migration of Buzzards is modest, numerically speaking, when compared to what happens in Falsterbö (south Sweden), where, and it varies from year to year, between 20,000 and 40,000 birds pass over. Even more pass via the Bosphorus with a maximum of 205,000 recorded in the last few years. On the other hand only 114 were counted in Messina in 2002. Ring readings of birds captured near Gibraltar show they come from Scandinavia, central Europe and Iberia, but the number of birds crossing is quite reduced. Quite a different matter is the huge number of Buzzards wintering in Iberia.

One difficulty to bear in mind when considering Buzzard migration is the great likeness between it and the Honey Buzzard, so if the counting is not done by expert ornithologists there may be errors in the resultant data. Even the most experienced birdwatcher in the Strait can mistake a distant flock of Buzzards for one of Honey Buzzards, especially when they are circling upwards in a thermal. But the Buzzard on postnuptial migration is much more of a soaring bird than the Honey Buzzard. As they approach the Strait they fly more inland, only visit the Rock sporadically and, when winds are strong, particularly the *levante*, they follow the same patterns of reversals, hold-ups, floating etc as those of Short-toed and Booted Eagles and Black Kites. Fortunately the peaks of both migrations don't coincide and this makes identification easier. Most Buzzards cross in October and, sometimes, more in the second than in the first half of the month, but at a time when there are very few ornithologists observing migration.

The wind doesn't seem to be a determining factor for passage because, on either of the two, Buzzards cross the Strait at the confluence of the two Mediterranean sectors, i.e. at the mouth of the river Guadalmesí, possibly as a result of their flying inland as they approach the Strait. In corroboration of this, the Migres Project in 1998 detected 10 Buzzards from the Cazalla observatory and 50 from the one at Algarrobo. There may be between 500 and 1000 Buzzards crossing via Gibraltar, but since Bernis reported in1980 there have been no further reliable data.

According to my experience and that of my friends in Gibraltar, over the last few years there has been a drastic reduction in numbers of Buzzards on postnuptial migration: No longer do you see the same size flocks as in the 70's and 80's. In fact, in some years you only see small groups. Paradoxically you still seem to see the same number of wintering birds in the area, which makes one think that migratory habits have changed and more are wintering in Europe possibly because of climate change – although these days this seems to be the panacea for everything for which we find no satisfactory explanation.

The Booted Eagle

The Booted Eagle (*Hieraaetus pennatus*) is a monotypical species widely distributed in the southern Palaearctic, in South Africa and Namibia (Kemp & Kemp, 1998) with two well-defined colour morphs, both pale and dark phases. There are an estimated 3,600 to 6,900 breeding pairs in Europe, of which more than half live in the Iberian Peninsula (BirdLife International/EBCC, 2000). It is absent from central Europe and Italy, but has reappeared in the Balkans (Purroy, 1997). It is a trans-Saharan migrant and is found in most parts of Spain, where it is only absent from the Galician coastal provinces, most of Catalunya, Almeria and part of the Eastern Mediterranean coast. It is sedentary in the Balearics and breeds in Mallorca and Menorca

(Viada, 1996). The Booted Eagle is essentially a forest raptor, which, like the Short-toed Eagle, needs large clearings near its breeding area. It lays two eggs and normally rears one chick; in many cases the other chick is killed by the older one. Iberian Booted Eagles on average have shorter wings (Tellería *et al., 1996)*. After three chicks were ringed in Cádiz and Huelva and recovered in Barcelona, Tarragona and northern Italy, between 23 and 159 days later it was concluded that they undertake long journeys northeastwards prior to migration (Tellería *et al.*, 1996).

Prenuptial migration takes place a little later than the Short-toed Eagle's and they normally appear at the end of February, a couple of weeks later than the Short-toed. Like the latter they don't travel in flocks and groups of them are even more scattered than those of the Short-toed, rarely do you see several flying together as sometimes

Booted Eagle circling upwards in a thermal. Thanks to its small size it is a very versatile soaring bird in thermals.

An Osprey on a cliff at Punta Carnero holding a freshly caught grey mullet in its claws.

occurs with the Short-toed Eagle, with which we are constantly comparing them as they have many points in common. I have seen flocks of Booted Eagles form and take a thermal together after the arrival of several individuals at the coast has coincided, but not because of a common flight strategy. Sometimes they tag on to groups of Black Kites or Honey Buzzards in a flock and when that happens the Booted Eagle is a dark phase one and the observer who is counting usually puts it down, erroneously, (and mostly) as a Black Kite or sometimes a Honey Buzzard.

You see them arriving at the European coast flying low when the wind is strong and letting themselves drift, although not so much as Black Kites do. They flap strenuously when crossing and don't do much soaring. One of the characteristics of this small eagle is that it often ruffles its feathers in mid-flight, normally when it has reached *terra firma*. It stops flapping, glides for a few metres and ruffles its feathers energetically; very often it lifts its head over one shoulder and has a look back. Within thermals its turns are small, as you would expect from an eagle of its size. The majority of them cross when winds are light to moderate, but some, although only a few, come over with strong prevailing winds.

On postnuptial migration their behaviour is similar to that of the Short-toed Eagle in that they are difficult to count, because they don't migrate in flocks and there is a lot of floating, reversals and hold-ups. The migration of the two species coincides both in time and space, so that you often see Booted Eagles flying in among very spread out groups of Short-toed Eagles. When floating they *suffer* more than the latter, due to their smaller size and longer, narrower wings, less adapted to soaring. Like Buzzards and Short-toed Eagles they approach the coast from the interior and cross at the confluence of the Alelíes and Tolmo-Faro sections (at the mouth of the river Guadalmesí). On very windy days, when they don't cross, you can see them in what look like flocks, but are really the result of a temporary bottleneck in the Strait. If a strong *levante* blows for a few days it is really an impressive sight to see such an unusual number of this species together. According to Paul Rocca, an ornithologist from Gibraltar, and Dr John Cortés, the author of Chapter III of this book, up to 600 individuals may form on the north side of the Rock and reverse towards the area around the mouth of the river Palmones.

Counting Booted Eagles is not easy and the figure of between 3,900 and 4,900 in the 70's, offered by Bernis, isn't exact, although he is the only one to try. Their passage via the Strait of Messina is accidental, like that of the Lesser Spotted Eagle via Gibraltar and via the Bosphorus the largest number recorded is 520.

The Osprey

The Osprey (*Pandion halkiaetus*) is a raptor with world-wide distribution and is found near the seacoast and large bodies of fresh water. There are an estimated 8,000 to 10,000 breeding pairs in Europe. In Spain the sub-species *P.h.Haliaetus* breeds in the Balearic Islands (Mallorca, Menorca and Cabrera), the Canaries (Tenerife, La Gomera, El Hierro, Lanzarte, Alegranza, Montaña Clara and Lobos) and in the Chafarinas (Melilla). In 2000 a pair built a nest at a reservoir in Bornos (Cádiz), not far from the Strait (M.Barcell & J.R. Benítez, personal communication). In 2005 two pairs tried to breed in the area but were unsuccessful. The Spanish population is estimated at 30-38 breeding pairs, none of which are on the mainland. On prenuptial migration these birds arrive in March and April, but it is not unusual to see individuals in May. They are flying at great altitudes when they arrive over the coast, flapping their powerful wings.

In March 1996 during the filming of a documentary called *The forest's heartbeat*, I was at the Strait collaborating in the filming of the arrival of Honey Buzzards, together with the photographer Antonio Sabater and the film director of the documentary Joaquín Gutiérrez Acha, both friends of mine. Suddenly we saw an Osprey perched on one of the rocks at the Punta Carnero lighthouse in Algeciras. As we approached it to take photographs it took off and we could see it was holding a fish in its claws, a grey mullet (*Liza sp.*) which it didn't let go of. This is the only evidence I have of this migratory bird perched on the cliffs of the Natural Park of the Strait, although I do have some from Las Lances beach. This anecdote serves to remind us that these birds see no obstacle in migrating over the sea, as they no doubt consider the sea itself an important source of food.

Some Ospreys from Scandinavia, the Baltic States and Scotland winter in Iberia and the majority are from Sweden, as is evident from the number of rings recovered (Bernis, 1980). You can often see up to half a dozen wintering on the rivers of the Campo de Gibraltar and in some years they have even wintered on the reservoirs. It is also quite common to see them fishing near Los Lances at the mouths of the rivers Jara, Vega and Salado. In Africa, according to satellite tracking results, some British birds winter below the western Sahara. Some birds have been tracked the relatively short distance to winter in Iberia or Morocco, but others go much further to Senegal. The most famous case is one we mentioned in Chapter IV of an Osprey which travelled 12,500 km from Finland to southern Africa.

They cross the Strait in any prevailing wind, although they do prefer westerlies. They approach the area via the Rock of Gibraltar and when the wind is an easterly they start at Tarifa possibly after following the coastline. Some arriving at Gibraltar cross from there, but others fly across the Bay of Gibraltar and cross to Morocco from the Mediterranean sector of the Natural Park of the Strait.

According to Bernis, the average number of birds crossing between 1972 and 1977 was 51 and the majority of them crossed between 9:30 a.m. and 15:30.

Kestrels

Although not strictly belonging to this chapter, i.e. the seasonal migration of soaring birds, we are going to refer briefly to some of the most important *Falconiformes* to be seen migrating across the Strait.

The *Falco* genus of raptors are small and medium-sized birds, expert flyers, whose strong wing beats give them great speed. They are recognizable by their pointed wingtips and most hunt live birds. Some of them, especially the Lesser Kestrel, but also the Common Kestrel and the Peregrine Falcon, breed in cities. They don't necessarily have to migrate across straits because, being such strong flyers, they can cross the Mediterranean or the Gulf of Cádiz anywhere and only a minority cross via the Strait of Gibraltar.

The Lesser Kestrel

The Lesser Kestrel (*Falco naumanni*) is a monotypical species distributed throughout the southern Palaearctic. Its wintering areas are not altogether well known but South Africa is ap-

The Male Lesser Kestrel, in full flight, is another migrant that doesn't necessarily prefer straits for migration purposes.

parently one of their preferred destinations. In Europe there are between 12,000 and 18,000 breeding pairs and they are more abundant in countries on the northern coasts of the Mediterranean in Spain, Italy, Greece and Turkey (BirdLife International/EBCC, 2000).

In Spain they are found more in the south west half, especially in Andalusia, Extremadura, Castilla-León and Castilla-La Mancha. There are an estimated 12,000 breeding pairs in Spain (Atienza *et al.*, 2001), and relatively stable numbers over the last few years, but with ups and downs in some areas. In the area of the Campo de Gibraltar there has been a sharp decline in the last few years. For example, in the city of Al-

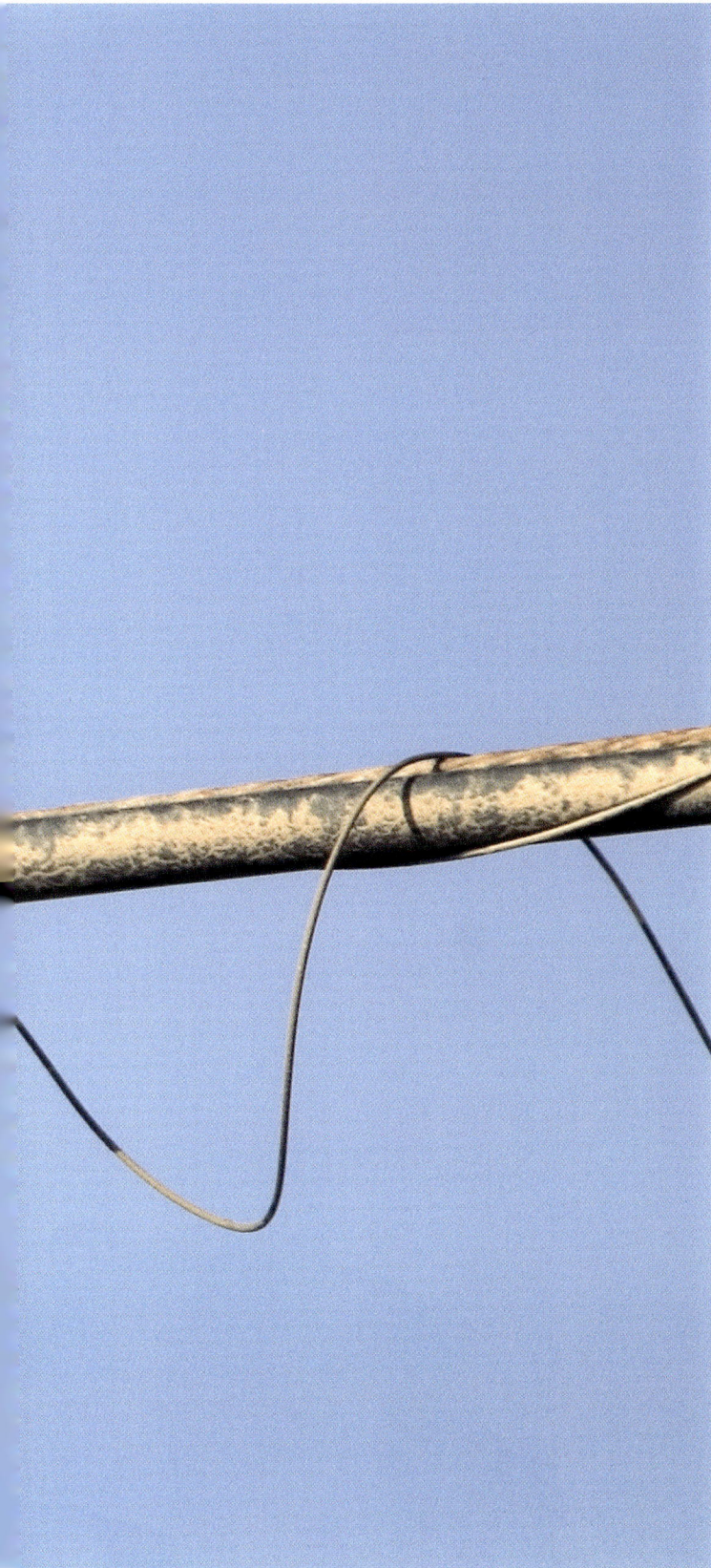

Male Kestrel perched on a cable near the Punta Carnero lighthouse, where this species has been nesting since the 1980's.

A female Kestrel soaring in a thermal near Algeciras.
With its feathers fully spread out it increases its body
surface area and is easily lifted by the dynamic thrust
in thermals.

geciras I counted 31 pairs in 1994 but only 3 pairs in 2004. The population on the European side of the Strait is largely sedentary and they spend the winter in colonies.

This little falcon breeds in reasonably large colonies on rocky cliffs and buildings (in small and large towns), which are surrounded by suitable hunting grounds such as waste land, fallow fields and stubble fields, where they find insects which form their staple diet. When rearing their chicks they tend to eat small birds and field mice.

On prenuptial migration they reach the city of Algeciras in the second half of January. This city looks onto the Strait and as I have lived here for many years I am able to monitor their arrival closely. Further inland in Andalusia their arrival has been recorded in February (Negro *et al.*, 1991). It is the second bird to return from its wintering sites, the first being the White Stork. The entry of these migrants is not detected for the reasons already mentioned. On occasions, well into February, I've seen one coming in over the sea very fast and not bothering with thermals at all. So this migration is not one that an ornithologist can really observe closely.

On postnuptial migration the countings carried out by Bernis are the most insubstantial of all migratory species, although there is another factor which makes it even more difficult to get reliable results: the similarity both in plumage and flight between the Lesser Kestrel and the Common Kestrel. Therefore the low numbers recorded in Gibraltar must be treated with caution.

The Common Kestrel

The Common Kestrel (*Falco tinunculus*) is distributed throughout Africa, Asia and Europe. Total population is between 300,000 and 440,000 breeding pairs (BirdLife International/

EBCC, 2000). The main species, of western Palaearctic distribution, is found in both Iberia and the Balearics. In the central and western Canary Islands there is a subspecies *F.t. canariensis* and in the remainder of the archipelago *F. t. dacotiae* (Martínez, 2003). The northern European populations are migratory whereas the Iberian populations are more sedentary. In Spain it is found everywhere, including the archipelagos, as well as Ceuta and Melilla, and there are an estimated 17,000 breeding pairs (Martínez, 2003), although with 17% of the territory still to be checked. It nests on rocks, ledges, trees and buildings near pastureland, meadows, etc.

Among the mainly sedentary Spanish population, a dispersal of first year kestrels, restricted to Iberia, has been detected (Tellería *et al.*, 1996). The Iberian population is joined in winter by a large number coming from north and central Europe, which enter the peninsula round both ends of the Pyrenees.

It is very difficult to quote reliable figures for the Lesser and Common Kestrel, due not only to the problem of distinguishing them in flight – which in itself is not as patent as that of soaring birds - and the fact that they don't need to migrate via the Strait, but also because there are good-sized populations in the area and it is easy to mistake their comings and goings with migrants. On the European coast of the Strait there are several pairs of Common Kestrels whose role in migration is an interesting one.

Other minor and occasional migrant raptors

In this section we refer to those birds of prey which have at some time been observed migrating across the Strait, some we are absolutely sure of and some not so sure. It is quite usual for ornithologist beginners to claim to have seen a rare bird and they don't hesitate to call a Red Kite what is really a young Black Kite, without bearing in mind that the obvious is the most important: the migratory phenomenon and the flight strategies they resort to in order to make a short, but, due to the strong year-round winds, dangerous sea crossing. If a Spanish Imperial Eagle crosses, then that, in itself, is of no great importance except for it being a beautiful species, very scarce and, therefore, occasional and it should arouse no further interest on the part of a good ornithologist. This is why those people who are in charge of migratory studies should advise their collaborators to be cautious in their observations, but above all not to be frivolous, because their results may be discredited by making unrealistic claims.

In 1972 when I was taking part in the first migration field studies to be made in the Strait, there used to be endless and sometimes very heated arguments between Professor Bernis and observers who claimed to have seen a bird which was hardly ever spotted crossing the Strait. It was not at all easy to convince him that what the person had seen was an immature Spanish Imperial Eagle and much harder still if it was a Spotted Eagle. It very often depended on the confidence he had in that person and their ornithological experience.

The Black-winged Kite

The Black-winged Kite (*Elanus caeruleus*) has recently been classed as a nocturnal raptor, due to the asymmetry of its ears and its habit of hunting at dusk. It has been breeding, in a minor way, in the area since 1992 and near the Strait there are about four pairs. On prenuptial migration I have seen it once, together with my late

A defiant-looking Black-winged Kite. This African raptor has recently colonized areas near the Strait.

The Black Vulture referred to in the text,
inside an old van waiting to be taken to
a recovery centre.

Basque friend Joseba Bernaola, coming in over the lighthouse at Punta Carnero in Algeciras on 6th April 1998. On postnuptial migration Bernis mentions it once, in 1997, and the Migres Programme three times in 1998 but there is no further data to know whether the bird was the same or not. In September 2005 a maximum of 27 individuals were counted at a roosting place at the former Janda lagoon.

The Black Vulture

The Black Vulture (*Aegypius monachus*) has been seen in the area several times both at the old rubbish dump in Tarifa and the new one at Los Barrios, always in the autumn-winter period. I have made several sightings at the Sierra de El Niño and one in particular, that I would like to tell you about.

One Sunday afternoon a friend of mine phoned to tell me that there was a Black Vulture in his garden at Punta San García. He told me they had seen it fall into the sea a few hundred metres from the shore in the Bay of Algeciras which his house overlooks. His son had got into a boat and gone out and rescued it. I said it was probably a Griffon Vulture because in the middle of the migrating season it wasn't unusual to find them in the sea or in the town, absolutely overcome with exhaustion. Anyway, I said I would go and collect what I expected to be a Griffon Vulture. When I arrived, what did I see but a magnificent juvenile Black Vulture and, as I had nowhere else to put it, I placed it in the boot of the car. I went to try and locate the warden of the then Pelayo Bird Recovery Centre, José María González, but he wasn't at home although I called there about three times. As I couldn't leave the poor bird in the car all night and the warden probably wouldn't be back until Monday morning, I decided to leave it in-

Rüppell's Vulture landing near a
carcass in the Santuario valley in Tarifa.

side an old abandoned van standing in the yard at the Centre, leaving a note for José María. The bird spent the night there and was found by the warden next day and what a surprise it was, because it's not every day you see a Black Vulture in your van! After it had recovered, several weeks later, a politician came to see it being released… I read about it in the local paper!

Rüppell's Vulture (Rüppell's Griffon Vulture)

Rüppell's Vulture (*Gyps rueppellii*) has been sighted in the area since the mid 90's. When the first ones were spotted it was thought they had escaped from a zoo but after several sightings in different parts of Iberia the suspicion arose that they might be migrant birds which had joined the returning Iberian vultures on migration. However, according to the Black Stork Ornithological Association (COCN) on 26th March 2003 an individual was seen arriving at Los Lances by Salvador Solís, who saw it being attacked by seagulls – a common sight in the Strait – before disappearing into the hinterland shortly afterwards. He estimated it to be a 2 or 3-year-old bird. On 25th March 2003 the same birdwatcher, together with other colleagues, saw another individual at Tahivilla (Tarifa), this time a one-year-old with about 70 Griffon Vultures, eating the remains of a dead cow. Finally, on 28th May 2003 another juvenile was captured near the N-340 road and from the look of its plumage it was probably the former of the two. It was taken to the zoo at Jerez de la Frontera, where it recovered and was released on 9th July 2003, after being fitted with a radio transmitter for monitoring and dyeing some of its primary feathers for ease of identification in flight. In 2005 at the Cazalla observatory I saw three of these vultures

A Rüppell's Vulture pictured arriving in Tarifa. This African vulture is becoming a more common sight in the area and in the rest of Spain.

soaring in a thermal with some Griffon Vultures and in September a young one was killed by the blades of a wind turbine at Tahivilla (Tarifa).

This habit of flying with Griffon Vultures makes me wonder if, just as they come over to Europe from Africa "dragged" along by the prenuptial migration of Griffon Vultures, the same thing happens to those present in this area and they go back to their sub-Saharan origins.

The Goshawk

The Goshawk (*Accipiter gentiles*) is classed as a sedentary bird in Iberia. It breeds in Los Alcornocales and, from the few sightings reported, one has to be sure it is a migrant bird and not a local one. I have only seen one arriving over the sea, when it came in on prenuptial passage, flying past one of the observers at waist height and on, low, over the hilltops. It cannot be mistaken for a Sparrowhawk, not only because of its size but because its flight is totally different. In the censuses conducted by Bernis in the Strait of Gibraltar there have never been more than a dozen and the Migres Programme has only recorded four sightings at one observatory.

The Long-legged Buzzard

There are two species of Long-legged Buzzard (*Buteo rufinus*): the one distributed throughout Central Europe and Asia Minor as far as Central Asia (*B.r. rufinus*) and the one in North Africa, (*B.r.cirtensis*) which is smaller and more of a pale red in colour. There have only been seven purported sightings in the Iberian Peninsula. I mention it here because, as it breeds in Morocco, it may be seen migrating. I am fortunate to have seen it on four occasions: two on prenuptial and two on postnuptial migration. On prenuptial passage I saw one in Gibraltar being harassed by several seagulls, and I saw another on the same day as I saw the Black-winged Kite with Joseba Bernaola. In August 2004, when I was on the hill at Cazalla with Joaquín Mazón, an ornithologist from Extremadura, and Ramón Sanz, a migration enthusiast from Catalunya, we had a clear sighting of a juvenile against a beautiful, blue sky.

The Rough-legged Buzzard

The Rough-legged Buzzard (*Buteo lagopus*) is found in the western Palaearctic and it winters in central Europe. It is easily confused with *Buteo buteo* and *Buteo rufinus*. On 31st October 1976 one was claimed to have been seen crossing to Morocco but this was later rejected by the Iberian Rare Bird Committee. Only one sighting has been officially accepted, on the Llobregat river delta in March 1990.

The Lesser Spotted Eagle

The Lesser Spotted Eagle (*Aquila pomarina*) breeds in Eastern Europe and Asia Minor, migrating mainly via the Bosphorus, where, in 1996, 18,898 were counted. It winters in Africa from the Nile valley down as far as Mozambique. In the Strait there have been 12 sightings corresponding to four birds (Barros & Ríos, 2002). One of these was seen by absolutely *all of us* because it stayed in the area from 7th to 23rd September 1998.

The Spotted Eagle

The Spotted Eagle (*Aquila clanga*) is found throughout the central Palaearctic. Although only an occasional visitor it has often been sighted in the Strait and both Bernis and observers in the Migres Programme have seen it several times.

The Spanish Imperial Eagle

The Spanish Imperial Eagle (*Aquila adalberti*) is only found in the Iberian Peninsula although some authors consider it to be a subspecies (Vaurie, 1965; Cramp, 1980). There were many sightings of juveniles and immature birds in the 70's recorded by Bernis (1980) and five sightings in the Migres programme of 1998. In 2004 we saw an immature bird from the Cazalla observatory, much to the excitement of all those present, and another in 2005.

Opposite

A year-old Spanish Imperial Eagle. They can often be seen on postnuptial migration near the Strait.

This Lanner was photographed on postnuptial migration preying on pigeons which nest in an old loft near the Comisario lagoon.

The Golden Eagle

The Golden Eagle (*Aquila chrysaetos*) is a relatively frequent winterer in the area round La Janda and sightings have been reported on several occasions both by Bernis and the Migres programme. On two occasions I have seen young ones set off on their Strait crossing.

The Hobby

The Hobby (*Falco subbuteo*) is a trans-Saharan migrant and seldom is it not seen several times on both migrations. I have always seen it flying very quickly and ornithologists must have their wits about them to catch a glimpse of it!

The Lanner

The Lanner (*Falco biarmicus*) is of North-west African distribution and is thought to have once bred in Iberia in the Guadalquivir marshes. There was an unsuccessful attempt to breed it in Menorca in 1971 (Muntaner & Congost, 1979). For my part, I had an excellent view of one for several days during August 2002 at the Comisario lagoon in Puerto Real.

The Peregrine Falcon

The Peregrine Falcon (*Falco peregrinus*) breeds on both sides of the Strait and it is not at all unusual to see pairs crossing from one side to the other to hunt. On several occasions I've seen one falcon over the sea chasing another one away towards its nesting area on the opposite shore.

Other large migrant birds

Here I am referring to large birds- the Greylag Goose, Crane and Flamingo - which are seen less often because their migrations don't coincide with those of soaring birds and are much more dispersed.

The Greylag Goose (*anser anser*) migrates at night and I have only heard them crossing over to Morocco once, in November. In February, in daylight, I've seen them flying northwards at altitude in their characteristic V-formation. In the 80's a small group wintered near La Janda but from the 90's onwards they haven't been seen again (Barros & Ríos, 2002). Flocks of Greylag Geese fly to the wetlands on the Moroccan coast.

Until it dried out in the 50's there was a native population of Cranes (*Grus grus*) at the La Janda lagoon, where they bred. Some claim it was a smaller subspecies. I have been observing them since the 70's and the population has varied from 600 to 1,300 individuals where the lagoon used to be. They also fly over to Morocco in autumn and I have been able to spot them many times because they tend to fly very high in flocks and emit a characteristic trumpeting call. Only once did I and some students and teachers from the Complutense University in Madrid see a group of a hundred or so flying low over the hill at Cazalla, late in the evening.

I've also seen Greater Flamingos (*Phoenicopteus ruber*) on two occasions flying over to Morocco during the day. These birds usually fly at night and you can only detect them by the calls they make when flying in groups. When I saw them they came flying along the edge of the Bay of Gibraltar at medium altitude, past the Punta Carnero lighthouse and further along the coast before gradually veering southwards. This migration takes them down along the Moroccan Atlantic coast to the wetlands of Senegal or over the interior when there has been abundant rainfall, enabling them to stop over at the famous dunes called Erg Chebbi, near the little village of Merzuga, where a shallow lagoon forms.

The migration of the
Greylag Goose is another
tremendous spectacle
which ornithologists
have to thank Nature for.

A grain in the balance will determine which individual shall live and which shall die, which variety or species shall increase in number and which shall decrease, or finally become extinct.

Charles Darwin
The origin of species, 1859

Misadventures
and deaths

Hunting

Hunting during the migrating season is by far the most common cause of death among migrating birds. The most notorious case is on the Malta archipelago, situated in the centre of the Mediterranean Sea, 89 km south of Sicily, where there exists an ancient tradition of shooting migrant birds and nowadays it is a prime example of the ideological madness of groups whose only argument is that they enjoy an anti-natural activity: that is, they kill animals for mere pleasure. According to a study by BirdLife Malta, every year about 5,000 birds of prey are shot, including 1,000 to 1,500 Common Kestrels, 500-600 Hobbys and a very sad, long list of others which we won't detail here.

Until the mid 1980's there was a very deep-rooted tradition in the Campo de Gibraltar, La Janda and in Andalusia in general, of hunting little birds with nets or simple traps. In the Campo de Gibraltar hunters would count what they had caught not in individual birds but in dozens, so great was the quantity. They used to catch most of these little birds on postnuptial migration by preparing the ground beforehand. As they knew from experience where the birds flew over they put sticks in the ground with a piece of cloth waving in the wind on top and these "directed" the birds towards mist nets where they were trapped. Another ploy was to place nets on the ground, operated with string by a poacher from a distance, and place "blind" birds on top of them to attract passing flocks and tempt them to land. These "blind" birds had had their eye pupils burned out with a red hot needle in the belief that they sang better. But the poachers' real favourite was the stick and trap because it's much easier to carry and less conspicuous than

A drowned Griffon Vulture. Another eleven were washed up on the beach. This is the first proof we have of the death of a flock of soaring birds during the Strait crossing.

the net. Sadly, these little birds were fried and eaten in many bars all over Spain and millions were hunted for this purpose.

In order to put an end to this practise, the most efficient method to be tried out entailed breaking the tradition. Thus, in the first year, the public authorities which regulated hunting (in the 1970's there was no Ministry of the Environment) issued a high number of licences to trap little birds and then each year gradually reduced the number a hunter could trap until, in Andalusia, the number was down to five birds per hunter. Gradually, the authorities thought, this most cruel form of hunting would become less popular because fines for hunting without a permit are high. So, if the father in a family didn't hunt, his son didn't become a hunter either. Poaching would always exist but it wouldn't be nearly as popular as it used to be. As it turned out, the result is not what it should be because checks are insufficient and so the cruel practice of hunting little birds still goes on. There are certainly fewer hunters now, but this savage custom still hasn't been eradicated.

Another very important cause of death is the loss of habitat. It may seem to be particularly grave in underdeveloped countries where extensive areas of land are converted into arable land, wetlands are drained etc, and it is precisely this type of land which migrants use as wintering sites. Paradoxically in the so-called first world too, land which was previously green belt suddenly becomes prime building land, wetlands are drained, ski slopes are widened and conservation areas, including national parks, are reduced in size. An in-depth analysis of this state of affairs is outside the scope of this book, which centres on the Strait and its area of influence, so we will examine the situation here.

Iberian migrant birds and those from beyond the Pyrenees arriving at the Strait on post-nuptial migration, most of them having gone several days without feeding, may be unlucky enough to coincide with one of those stormy *levante* wind periods. This will force them to wait a few days before being able to cross the stretch of sea and continue their journey on to dry, sandy sub-Saharan Africa or to the damp jungle of central Africa. Unfortunately, at about this time in August the hunting season begins for some animals and there are, therefore, hunters in the countryside – although hunter in this case is a euphemism for trigger happy people who hide behind their hunting licence to shoot protected species. Genuine hunters are those who respect the law and only shoot authorized species.

We don't know for sure how many raptors die in the area, victims of human ignorance, but, fortunately, the time has passed when some gunmen would boast in public about shooting "eagles"; nevertheless such cruelty will always exist. There is only one way of knowing the quality of hunting etiquette in the area and that is by consulting the records at CREAS (Reception Centre for Injured Animals) in the Campo de Gibraltar or the provincial centre, where you can find out the origin of injured birds, dates and other information.

Power lines and wind farms

In spite of what I have already said about hunting, I am convinced that in the Strait the prime cause of death among migrant birds is not hunters but electricity power lines, which, like a poorly planned and even worse situated spider's web, cover a very wide area. In the area there is a coal-fired power station, another one that is gas

fired and at least three combined gas/coal which distribute their production via transmission lines. Furthermore, let's not forget the 400,000 volt undersea lines which take electricity to Morocco and the burgeoning wind power industry which already produces 150 megawatts. All this is very, very near to the coast of the Strait, which an estimated 300 million birds fly over every year on both migrations, according to a recent study conducted by the Max Planck Institute at Radolfzell in southern Germany.

Up to the end of 2004 only one study had been carried out into the impact of wind farms on medium to large-size migrant birds. This was published on 5th June 1995 and, despite pressure from local ecologist groups and sheer common sense, there has been no follow-up study, a fact much criticized on the Internet by one Mark Duchamp in Ibérica 2000. The study lasted a whole year and detected the deaths of 30 Griffon Vultures, 12 Common Kestrels, 3 Lesser Kestrels, 2 Short-toed Eagles, 2 Eagle Owls, 1

Honey Buzzard, feathers damaged by a hunter's shotgun pellets. The raptor made a huge effort to continue its migratory journey.

Black Kite, 1 Black Vulture etc, 69 raptors in all - and bear in mind that only 34% of the wind turbines existing at the time were monitored. If we extrapolated these results, even though it may not be "politically correct", we would get 90 Griffon Vultures a year, 30 Common Kestrels etc. In other words, during these past ten years 900 Griffon Vultures will have died.(And this doesn't take into account the increase in the number of wind farms) But the problem is not just the turbine blades: the evil lies also in the power lines that birds can neither see nor hear when flying over the area, although the study concludes that, of the 106 dead birds, 97 were found under wind turbines and 9 under power lines, but without stating how many kilometres of power lines were examined. It says that 93.33% of vultures were found dead near PESUR (a wind energy company) turbines, but since then, as far as we know, no action has been taken by either PESUR to improve these poor statistics, or by the Andalusian Regional Government, which asked SEO/BirdLife to carry out the study.

The deaths of these birds always occur in the secluded countryside with no prying witnesses and to discover or study this phenomenon requires a considerable financial outlay for anyone interested; so unless a public or private company puts up the money we shall continue to be in the dark about these dramatic numbers of victims. Indeed, this reluctance to carry out field studies is as if we are accepting all we fear to be true as a foregone conclusion, and, what is even worse, this means we accept the so-called ecological tax which nature has to pay so that we humans can live a "better" life. But if, as the SEO study shows, there are certain "black spots" at some wind turbines and power lines then they could be corrected at little extra cost and this

A Griffon Vulture, blown by strong gusts of wind, crashing into power lines.

only goes to show up the indolence and indifference on the part of the companies involved and the Regional Government and, why not say so, local ecologist groups whose inaction benefits unscrupulous commercial interests.

In this section we must also mention those deaths produced by collisions with vehicles, although those birds involved tend to be almost exclusively Passeriformes and medium-sized birds, not soaring birds. But, as always, there are exceptions.

After the Spanish government introduced some timid regulations on the wind farm industry on the Spanish coast of the Strait, many companies, whose sole concern are their balance sheets, decided to set up wind farms on the more permissive Moroccan side of the Strait, which is now being invaded by these "murderous windmills". Of course, any sensible person with a basic knowledge of wind energy will not be against it but what we do object to is this form of energy at whatever the price and maximum profitability, cost what it may, in this case the birds, which cannot complain anyway.

However, within this bleak picture, there is an island of common sense and a company which thinks, logically, about its balance sheet but also about nature and its conservation, making income compatible with the survival of migrant birds.

In September 1998 Desarrollos Eólicos S.A., a wind energy company with 100 turbines - called TA-1 as it is the first wind farm to be situated in the municipality of Tarifa - started operating in the Armarchal, Zarzuela and Tahivilla area, where free range cattle farming on the hills is an important part of the local economy. During the first two years no measures were taken to prevent Griffon Vultures crashing into the turbine blades. This involved about 30 birds each year due to the nearby presence of a breeding colony at Laja de Ciscar, where 90 breeding pairs nest and roam the area in search of dead cattle.

Once they were aware of the problem the first step they took was to employ a guard, in permanent radio contact with the company's head office which would stop the turbines if there was imminent danger of vultures colliding with them. The number of victims gradually fell, especially when an additional guard was placed on weekend duty. Thanks to this measure the death rate has been cut by 70-80% and now only 6 or 7 Griffon Vultures die each year. However, according to one of the guards at this wind farm, Manuel Lobón, other species have also fallen victim to the turbine blades: the Common and Lesser Kestrel, Short-toed Eagle, Black Kite, Egyptian Vulture, Eagle Owl, Cattle Egret, Little Owl and an unknown number of *Passeriformes*.

Endogenous causes of death

Apart from the above-mentioned causes, to which most deaths can be attributed, there are others that we are going to refer to as endogenous in some migrant birds, such as the wrong "transmission" of the gene or genes which cause a bird to migrate. When this occurs it makes a bird that is migrating alone lose its way and not reach its destination.

Little or nothing is known about the toxic substances which make their way into the food chain of birds, including migrant birds, and what effect they may have on genetic transmission. It may be that malformation, which is easily noted in large birds such as the White Stork, is the least important problem. Possibly the most lethal effects are those which the human eye cannot perceive.

Oppositte

Short-toed Eagle struck by the blade of a wind turbine. Its right wing, seen in the background, was sliced off.

Shortage of food

The extreme weakening of a bird due to lack of food may occasionally be a cause of death. In this area we have often read in the local paper about Griffon Vultures falling into school yards, onto roofs, roads, factories etc. More often than not this happens in autumn and early winter, which is their postnuptial migration time. Among the more than two thousand vultures that come to the area, many of them are young and inexperienced and absolutely exhausted after going days without food so that they can hardly keep their wings spread. According to Kuroda the pectoral muscle of soaring birds is made up of two parts, a large upper part and a smaller deeper one whose volume, depending on the species, is between 8 and 11% of the whole muscle. The latter, when taut, is the one which keeps the wings spread while the bird is soaring. But when the vulture runs out of steam this muscle loses its tone and the bird is unable to keep its wings outspread, falling down in the most unusual places. The problem for these tired animals is that they have no control over where they fall and they will be lucky if they're rescued. Over the last few years, however, this situation has changed for the better and, although exhausted vultures still appear, it is unusual to read about a fallen vulture in the press. This is due exclusively to their dependence on the rubbish dump at Los Barrios, because on arrival in the area, possibly with others of their own species , they are attracted by other vultures round the dump and go there to feed.

In 2005 two feeding places have been set up in the Los Alcornocales Natural Park, one in the northern and one in the southern part, with two purposes. Because of mad cow disease it was forbidden to leave cattle carcasses in the fields, thus depriving raptors of carrion. These feeding places now provide carrion which has been given a safety check by the Andalusian government and helps both the local vulture colony and migrant birds. In addition they play an important role in this area and in Andalusia in the survival of the Egyptian Vulture, whose population is at a critical stage because of illegal poisoning. It is hoped that they will get used to these feeding places and numbers will then recover.

A study carried out at this dump by José Rafael Garrido, Cristina García Sarasa and Manuel Fernández Cruz (2002) found that the main food there consisted of   the remains of birds, fish, sea food, meat and bones. At the old dump in Tarifa I remember seeing several vultures eat a whole box of prawns, probably thrown away for being unfit for human consumption! At the Los Barrios dump there were fewer than 100 vultures in summer and spring and about 700 in autumn and winter, with a maximum of 476 seen feeding together. This study established that, although the dump is a supplementary source of food for vultures, it can also be a source of disease and poisoning. Juveniles will prefer to eat there because it is easier than searching for carrion, which is available in the area though. Somewhat paradoxically, because of the geographical location of the dump at the entrance to the Strait of Gibraltar, it increases the chances of survival of migrants, but many Iberian and French juveniles winter nearby and therefore don't migrate.

On prenuptial migration the vultures that arrive at the Moroccan coast are birds that, although immature from the photographs I've seen, are experienced in searching for food in African countries, where there is stiff competition and extreme conditions. So when they arrive they have considerable experience at gliding, but not at wing flapping, which wasn't

essential to their way of life. After their "African adventure" they come to an area where food is scarce (remember that the Griffon Vulture has officially been declared extinct in Morocco) and there is no dump at which they can "refuel" as they can at the other side of the Strait. Crossing the Strait is a terrific obstacle, the last one on their migratory journey that they must over-come to reach their breeding colonies in Iberia and France. Although we only have news of 11 drowned vultures found on Valdevaqueros beach

Juvenile Griffon Vulture, too tired to take off, now trying to recover at a factory near the Bay before continuing its migratory journey.

Juvenile Griffon Vulture on postnuptial migration, totally spent and hungry, observing a lady passing by.

in June 2005, many birds, weak through lack of food, must surely perish because they haven't the strength to keep flapping their wings throughout the crossing and, while flying and soaring, lose height and fall into the sea where they usually die from hypothermia. In 1994 students and teachers from the Complutense University of Madrid watched helplessly as a vulture fell into the sea, just 20 metres from the shore and died, trapped between two rocks, semi-submerged by the high tide. In an earlier chapter we have mentioned how many of these birds collapse onto the shore, unable to fly on and, even with the help of thermals, with no strength left to keep their wings outstretched.

Against all human logic, vultures arriving on prenuptial passage in the area hardly pause at the rubbish dump, as is evident from the above-mentioned figures. In spite of the tiredness or exhaustion in a large number of them only a few stop to store up energy. This behaviour is quite opposed to that of other migrants such as the White Stork or the Black Kite, which pack the dumps near the Strait during the spring migration, eager to restore energy used up on their journey. One might think that the birds are in a hurry to get to their breeding areas but this is not the case of Griffon Vultures which lay eggs in December and by April all the chicks in the colonies have hatched. So why the vultures are in such a hurry we just don't know!

Death by misadventure or lack of food isn't as frequent in other soaring birds as it is in vultures, possibly because their smaller size doesn't make them stand out, or because they have more fat reserves – e.g. honey Buzzards – or because some of them eat more often during migration, as *falconiformes* do. Storks, Black Kites, Ospreys and Booted Eagles are wont to hunt and feed while migrating, whereas Harriers, Short-toed Eagles and especially Honey Buzzards don't feed at all.

Misadventures or deaths caused by predators or attacks

As a general rule large birds are certainly not targets for other predators. On their migratory journeys they are in greater danger from hunters than predators and smaller raptors must always be on the alert because these little hunters can soon become the hunted!

Earlier, in Chapter II, I mentioned that a pair of Eagle Owls preyed on some migrant birds and how "*when monitoring a pair rearing two chicks in 1987 they preyed on the following species: Short-toed Eagle, Buzzard, Honey Buzzard, Barn Owl, Scops Owl, White Stork (chicks)…*". With the exception of the Barn Owl and the White Stork chicks, all the other species were captured while migrating. From this we can deduce that while this species is rearing its young, almost any of the migrant birds that are unfortunate enough to rest in its territory is liable to end up as part of its varied diet at this time.

Other dangerous predators for migrant birds are Peregrine Falcons which nest near the Strait and bring forward their breeding periods to coincide with the arrival of migrants. They don't hesitate to attack any bird, regardless of size, that encroaches on their territory, either to add it to their diet or to chase it away. If these attacks happen over the sea, as in the case of the Peregrine Falcons in Gibraltar, many migrant birds may fall into the sea, even one as big as a Griffon Vulture.

My experience of aggressive birds in the Strait is with Gulls and Common Kestrels during prenuptial migration. The Herring Gull

Eagle Owl`s nest. The female has a Honey Buzzard`s leg in her claws and spread around the nest are the remains of a Short-toed Eagle, rats, hedgehogs etc.

(*Larus argentatus*) is, to put it one way, more of a sea bird than the Yellow-legged Gull (*Larus michaellis*), at least during the breeding period. It patrols up and down the coast continuously and doesn't venture into towns like the Yellow-legged Gull does. Its small breeding colony is on the island of Las Palomas, not far from the Punta Carnero lighthouse and this stretch of coast is its territory. Thus, during the breeding period, any bird which crosses its path may be a target of its intolerant and irritable temperament. The largest breeding colony of the Yellow-legged Gull is to be found on the east face of the Rock of Gibraltar and, up to a few years ago several hundred breeding pairs of Herring Gulls nested in the Tajo de Barbate,.

On one occasion I saw four Herring Gulls chasing a Sparrowhawk just a couple of metres above the sea. The poor bird was at the end of its strength and the gulls were taking it in turns to tire the little raptor out, screaming at it non-

My friend Pablo Ortega, taking care of a Booted Eagle rescued from drowning in the sea after being attacked by seagulls.

stop to intimidate it before it could reach the relative safety of land. One that got nearest to it pulled out a tail feather, but the hawk managed to shake it off and I lost sight of them all as they flew at great speed round a cliff jutting out into the sea.

These aggressive birds take special delight in preying on weakened birds, no matter what their size. I have seen this happen many times: Short-toed and Booted Eagles, Black Kites, Buzzards and Griffon Vultures have all been forced down into the sea. An ornithologist friend of mine, Pablo Ortega, and I succeeded once in rescuing a Booted Eagle which had been forced into the sea by gulls. As luck would have it, the wind was a westerly and the waves brought it to the shore. We picked it up and took it to the Recovery Centre near Algeciras, where it was fed for a couple of days and then released without further ado.

When gulls spot a weakened bird they start squawking, probably to warn other gulls so as to harass it together. They fly directly above it and then, screaming away, they dive down to attack it with their beaks and sometimes even defecate on it in an attempt to dirty its feathers and make it heavier and slow it down. Every time a gull or gulls fly past a raptor they have singled out, the latter, in its struggle to dodge them, drops down a little from the route it was flying on. So, gradually, after each attack, it loses more and more height until the time comes when, in a last desperate attempt to dodge away from the gulls, which it is keeping an eye on more than the sea, it falls into the waves. After that happens, the gulls fly past a couple of times, but no more, and just fly away, leaving it to its fate. At that time of year, if the ornithologist is on the alert, they can foresee an attack because this in-

Kestrel attacking a Booted Eagle. These small falcons show zero tolerance to any migrating bird of prey which may venture into their territory.

A Barn Owl knocked down by a vehicle on the N-340 road just after making a successful crossing of the perilous Strait.

variably takes place when gulls spot a bird with damaged feathers or note from the way it is flying that it is tired or nearly exhausted. Short-toed Eagles and Griffon Vultures are, because of their heaviness, more vulnerable and they are the ones I've seen attacked most during prenuptial migration.

On the European coast of the Strait there are several nesting pairs of Common Kestrels, small birds of prey but very irritable during the breeding season. When migrants invade their territory they don't hesitate to attack them furiously, although the attacks I've seen didn't have serious consequences. Their breeding territory is smaller than that of the Herring Gull and, until a migrant comes very close to their nest they don't make a determined move to chase them away. They will have a go at Short-toed Eagles, Black Kites and Griffon Vultures, which put up no opposition, but the Booted Eagle will show them its claws as a deterrent.

All this action can only be observed on prenuptial passage at the coast and that is why, among other things, I prefer prenuptial migration, from a photographer's point of view. I am very lucky to be able to witness such incidents, sitting, so to speak, in the front row of the migratory show and watching these skirmishes. It is only possible if one understands the "small print" of migration in the Strait!

Poisoning

The rubbish dumps at Los Barrios and Medina Sidonia are two very important poles of attraction for migrant birds on both passages. I have mentioned earlier about how important they are for Black Kites, Griffon Vultures and White Storks, which are, by far, the migrant birds that exploit these sources of nutrition. The major

one for Iberian storks is Miramundo and, for storks breeding outside Iberia, the one at Los Barrios. At these dumps the storks fight with each other over all kinds of unimaginable objects seen by them to be tasty morsels. I once saw two young storks start to swallow a piece of thick string, from opposite ends, until their beaks touched and then after a considerable struggle one of them went off with the trophy. It's also common to see birds flying with a plastic bag attached to a foot, having been unable to shake it off. Scavenging among bags of rubbish they will swallow the most incredible things as quickly as possible so that their fellow scavengers don't snatch them away. The sight of these birds drinking in a black stream coming from the dumps is really obnoxious. The dumps don't only attract White Storks – in fact because of the dirt, some of them look like Black Storks – there are also tremendous numbers of Yellow-legged Gulls, Cattle Egrets, Jackdaws and, although fewer in number, different species of small birds. The storks that feed at the Miramundo dump go on to the Comisario lagoon to drink and wash their feathers, which get filthy among the greasy rubbish bags. Some of those which have eaten toxic substances die at this lagoon. Certain conservationist groups claim that there is a hygiene problem at the lagoon, when really this problem is solely at the dump. In my opinion these deaths are nothing to worry about as the figures involved are not at all alarming and, if it weren't for the dumps, many, many, more migrant birds wouldn't be able to tackle postnuptial migration successfully and, possibly, in some cases they wouldn't be able to breed for lack of food for their chicks.

Another cause of death among migrants, in this case sea birds, is the spillage of crude

Dead White Storks at the Comisario lagoon, possibly due to eating contaminated food at the Miramundo refuse dump.

Seagull killed by marine pollution
after an oil spill from a tanker in the
Bay of Gibraltar.

oil transported by ships in the area. About 800 ships a day pass through the Strait of Gibraltar: that is about 300,000 a year and a high percentage of them are tankers containing dangerous substances. Many of them oil tankers heading for the oil refinery which is situated right in the centre of the Bay between the towns of La Línea de la Concepción and Palmones, near los Barrios. There is a high risk of spillage and consequently the risk to migrants is also high.

Contact with spillage stains the birds' feathers and they lose their insulating properties, while at the same time they breathe in the toxic gases of these products and die from hypothermia or poisoning.

Deaths due to the wind

We have commented in previous chapters how the wind in the Strait of Gibraltar is strong and continuous and these are two dangerous characteristics for migratory birds crossing twice yearly. Furthermore, the direction it blows in is usually sideways on to the north-south and south-north route taken by the birds.

In chapter IV we looked at the serious problems posed by strong crosswinds, especially the *levante* winds when drift can take the birds away from the coast and into the Atlantic Sea, where there is no hope of finding land in order to rest. This is why the migrant birds crossing via Gibraltar fear these winds and normally don't risk crossing. With *poniente* winds of similar strength more birds cross than with a *levante*. However, in spite of this hereditary fear many less expert and impetuous birds do risk the crossing and are probably lost in the anonymity of the waters of the Atlantic.

One of the main features of the winds in the Strait is their habit of blowing in gusts, in other words their intensity varies in a short space of time, and they immediately recover what is lost or lose what is gained. When vultures crash into the wind turbines at Tarifa, the reason is mostly because gusts of wind destabilize young, inexperienced Griffon Vultures near the turbine blades. Nearly all dead ones found after colliding with blades are juveniles.

On postnuptial migration with a prevailing westerly wind I have seen some flocks of Griffon Vultures approaching a row of wind turbines on the Sierra del Cabrito, near the coast, and been able to observe the behaviour of these large birds in adverse circumstances. Two of them found themselves unexpectedly so near the blades that they partially folded their wings and stuck out their claws in a desperate attempt to avoid crashing and then descended to the ground just a few metres away from the deadly blades. Others were so close that they decided to fly between two turbines and that day they got away with it. Finally, still more saw the danger from a distance away and had time to gain height and avoid them completely.

We will never know the scandalous figures for migrant bird deaths due to man's negative impacts on the environment but we must find out what these are in order to take steps to minimize them, assuming that their total elimination is impossible. Straits are bottle necks where birds have to wait for several days and the more they have to wait the more danger there is of having misadventures, especially if such traps or dangers take the form of invisible power lines, wind turbines in the wrong places, unscrupulous hunters all lying in wait for them and let's not forget Nature itself, which has been sorting the weak from the strong ever since animal species existed.

An exhausted Griffon Vulture, quite famished after going without food for several days, resting on a beach on the north coast of the Strait on prenuptial migration.

There is no place anywhere in Spain where you can acquire knowledge of European raptors as quickly and abundantly as you can in Gibraltar.

Francisco Bernis
Bird migration in the Strait of Gibraltar.
Vol. I Soaring birds, 1978

Basic observation equipment: binoculars and telescopes

Human eyesight is not sufficient if we want to observe birds in flight from a distance, so we need optical instruments to compensate, in the form of telescopes and binoculars.

In spite of using optical equipment with lesser or greater magnification we mustn't forget that we humans "see" with our brain because we interpret reality, something which animals don't do and it could be that with binoculars we are observing things removed from reality. As an example I am going to tell you a true story that happened to my friend Cristina Parkes (Cristi), an ornithologist I've mentioned earlier.

At the end of the 80's Cristi got a letter from an important British ornithologist saying he would like to see Great Bustards (*Otis tarda*) and asking her if she would help him because he knew that around that time there were quite a lot at La Janda, not far from the Campo de Gibraltar. After an exchange of correspondence (in the 80's there was no Internet), Cristi picked him up from Málaga airport and drove him to her home near Algeciras.

The next day they left in the morning and drove to the former lagoon of La Janda surrounded by meadows and rolling hills, where they tried to locate the majestic bustards. After a few minutes Cristi spotted a group of about ten in a sown field. From their position they had a magnificent view of the birds and they could even make out the spectacular "moustaches" of the males, a feature of this species. The gentleman was really thrilled and couldn't thank Cristi enough for the big favour she had done him, so grateful was he at being able to see these enormous birds so close up.

Ornithologists from different countries get together at the Algarrobo observatory in Algeciras for postnuptial migration.

Although they had an excellent view, Cristi suggested going even nearer by driving round and stopping near some cars parked next to the field where the bustards were. Once they got there they crept up quietly behind the cars and when they looked through their binoculars they saw to their horror that the supposed bustards were really a group of farm labourers bent over working in the field, and the nearby cars were theirs!

The British ornithologist didn't say another word and took the first available plane from Málaga without uttering a word of anger or annoyance. He must have thought that my friend Cristi had played a practical joke on him, although she did tell me, half laughing, half sorry, that they had definitely seen the moustaches of the male "bustards"!

What sets the Strait of Gibraltar apart from other places, and is an important determining factor when choosing a pair of binoculars, is the wind. The wind moves absolutely everything, including our own body and will buffet us as we try to follow one or several birds with binoculars in our hands. If the wind is force 4 and you don't have something to rest on, you will have difficulties in seeing birds in flight clearly, and in a force 5 wind, even when resting on something, birds are very difficult to identify from a distance. In a force 6 and greater, there will be no ornithologists at the observatories and no birds to be seen! So the greater the magnification, the less we will be able to see clearly and in detail under these conditions. These days, binoculars are made up of sets of lenses and prisms. I assume that those of you who have acquired this book are more than familiar with these optical instruments and that is why I shall only refer to their use and not recommend any particular model.

Marta and Judith, ornithologists from the Vertebrates Department of the Complutense University in Madrid, scanning the sky for a flock of White Storks.

What is the ideal magnification for watching migrating birds? There must be a wide variety of opinions about this, all perfectly valid, but those of us who have had to put up with strong winds in the Strait will agree that large magnifications make the buffeting caused by the wind even worse and both vision and image quality suffer; so we would recommend 8x and maximum 10x magnification. Another important thing to consider is whether they should be light or heavy. Personally, to follow migration in very windy places, I recommend heavy binoculars that the wind will move less. If you're holding something heavy, your arms won't move about so much as they do when you're holding something light. I

Opposite
Joaquín Mazón, a specialist from the Migres Foundation, trying to locate Egyptian Vultures from the Cazalla observatory in summer 2005.

Time for a rest at the Cazalla observatory after watching birds passing over for several hours.

always used to have heavy binoculars and, on very windy days now that I have smaller ones, I long for my old ones. But, like everything else, it's a question of taste.

Over the last few years some binoculars have appeared on the market with an image stabilization system. I think these are ideal because they have a 12x magnification and the added advantage of making blurred images disappear at the press of a button: quite a luxury that an ornithologist can afford, as the price-quality ratio is excellent.

Terrestrial telescopes are optical instruments with one eyepiece whose main characteristic is the higher levels of magnification it can achieve. It consists of two parts: the telescope itself with a set of lenses, and the eyepiece. A high quality telescope has lenses with different types of glass to overcome what is called a "false colour" problem, and also less distortion and a minimum focal length, which can be as short as 5 metres. The eyepiece incorporates a zoom system of 20x – 60x, which is the most common, or there is a fixed magnification of 20x or 30x. The main disadvantage of telescopes is that they need to be on a tripod to avoid vibrations, which at these magnification levels make good vision impossible.

Which system is the most adequate for observing migration in the Strait? The answer to this question depends on what our intentions and needs are. If you are studying migration and counting migrant birds, then 8x30 or even 10 x40 binoculars are recommended. If you want to enjoy watching birds in flight, then use a telescope with a 20x60 zoom. Many ornithologists carry both instruments with them; a telescope for examining distant objects and binoculars when a flock or individual bird is detected.

Everyone knows about the loss of perspective when using a telephoto lens or other optical instruments with large magnifications. Thus, when watching a Formula I race head on, it looks as if the cars aren't going fast, or in a cycle race it looks as if the *peloton* is bunched together. The same thing happens with binoculars or telescopes which "squash" the perspective and, in our case, birds appear to be stuck to the background behind them, seemingly much further away than they really are. This miscalculation of a distance can lead us to think that a flock of storks is fly-ing over the middle of the Strait when in fact the birds haven't left the coastline.

How can this influence the observation of migrant birds? First of all it will mislead the observer into thinking that a bird or flock is much further away than it really is. If you see a flock, which is apparently a good distance away, then you may suddenly be surprised to find it is flying overhead a few moments later. If you tell other ornithologists at a different observatory to look at the same birds and give them their position, the position you give may be wrong. When giving the position of a bird or birds you have to speak about projection, by which we mean the position of a bird on an imaginary line joining our position with the background against which the bird is flying.

A new form of photography, born in the digital era, consists of combining a digital camera with a telescope. Both should be top quality articles, otherwise the pictures will be mediocre, even if they look good on a PC screen. The result is some fantastic magnifications and at the moment the only drawback is that digital files are rather small to be able to make medium-sized enlargements on paper, but the industry is always looking for new customers and will no doubt give them what they are looking for. So it's only a question of time before digiscoping fulfills the dream of every Nature photographer: to be able to photograph animals from a great distance so that they are relaxed and behaving naturally.

Identifying the most common soaring birds in the Strait

Once you have a pair of binoculars, or even a telescope, you are ready to observe birds at rest or in flight, but, although it's very easy to distinguish between a White Stork and a Black Kite, it's not so easy to tell the difference between a Honey Buzzard and a Common Buzzard from a distance on a misty day. In order to do so, we're going to give a few tips on how to identify birds by species and then continue by comparing those liable to cause confusion, while referring only to those seen most frequently in the Strait. To identify accidental birds we shall refer you to the specialized publications mentioned in the Bibliography.

Because of the meteorological conditions the Strait isn't always the ideal place to observe birds, even though that may seem a contradiction in terms. When strong winds are blowing they can destabilize observers and prevent them from enjoying a good view of birds. They also upset the birds which are forced into sudden turns, or rise and fall very quickly. Another frequent problem is the haze which makes it difficult to see birds arriving from a good distance away. The strong light and, what is even worse, the abundance of cloudy days, always turn bird silhouettes into black against white.

It may well be that you, the reader at home, who has been patient enough to read this far – and my sincere congratulations if you have – might be feeling that the Strait is an evil place, where strong winds take away naughty children and carefully groomed hair lasts less time in place than it took to get it to stay there. But it's not like that and, if I've given that impression in the preceding pages then I can assure you that it's the wrong one! Expressed as a percentage, there are many fewer "evil" days, but you tend to remember them more and that's where the difference lies. I must also point out that, to offset this, these evil days of terrible winds force soaring birds to fly at lower altitudes and this means you have the chance to see Honey Buzzards, Short-toed and Booted Eagles, Black Storks etc. at unusually low heights, all the better for the birdwatcher and, of course, the photographer to take advantage of.

Honey Buzzard *Pernis apivorus*

The Honey Buzzard's feathers nearest to its eyes act like scales to protect it from wasp stings.

The Honey Buzzard in Spain is referred to colloquially among local ornithologists and Nature photographers as a "pernis" which comes from the species' Latin name. It is a medium-sized raptor and usually travels in smaller flocks than the White Stork, but, together with the Black Kite it forms the most compact of all raptor flocks. It may be mistaken for a Buzzard because of its rounded wings and very visible dark carpal patches, but its head is less robust, its neck sticks out from its body more than the Common Buzzard's and its tail is longer and more rounded when circling in thermals. Adults can be distinguished from juveniles by the black band of trailing edge feathers and the striped underwing coverts and tail feathers. When circling in a thermal it spreads its wings straight out, while the Common Buzzard keeps them in a very open v-shape or obtuse angle. It flaps its wings in a deep, wave-like movement, after which it will glide with its wings kept horizontal. The impression it gives is that of a strong, determined bird, unafraid of adverse meteorology whether it be wind or rain. It displays a high degree of self-confidence and decisiveness when crossing the Strait.

Apart from the plumage phases you can tell adults from juveniles by their dark "fingers" and the black subterminal band in the trailing edge wing and tail feathers. I can strongly recommend a magnificent book by Dick Forsman, a regular visitor to the Strait, called "Raptors of Europe and the Middle East", to those who would like more information on this subject.

In the Strait it is easy to observe the passage of the Black Stork if you're there at the right time.

The Black Stork is a large bird, identical in size to the White Stork, but its white belly feathers, seen against the sky, almost form a triangle. Generally it is more black than white and when circling in thermals its upper feathers are totally black. Flocks generally contain fewer birds than flocks of White Storks but this is not a determining factor in identification. Flocks of several hundred have been reported but the average may be around 60. When speaking of White and Black Storks I usually refer to flocks rather than individual birds because you don't often see a stork on its own, although if you do it is likely to be a Black rather than a White Stork.

White Stork *Ciconia ciconia*

The White Stork is a large bird, easily identified by its black and white underparts, stretched out neck and feet, and its red bill and feet. If the birds are far off and it's a bit hazy we may be in doubt about whether they're White or Black Storks, but if we look at their silhouette carefully when they are circling in a thermal, we can spot an important detail; the upper part in flight is in general white, although there are a lot of black feathers too. There is only one other bird, whose upper feathers are so white, that could be confused with a White Stork, and that's the adult Egyptian Vulture, although it's smaller and doesn't have long feet or neck. Storks usually fly in big, compact flocks, the largest ones of all those migrants passing over Gibraltar. On a day of clear, blue skies it's a spectacular sight to see several hundred birds circling upwards together, very close and sometimes you see their underparts and sometimes their back, so that the flock at first appears to be more white than black and then more black than white. On postnuptial migration you may see several thousand in a flock, although the average flock size is about 400 to 500.

The White Stork is one of the most numerous soaring birds and is easy to identify.

When the budding ornithologist spots a bird, the first thing he must ask himself is what its size is: large, large-medium, medium, medium-small and small. As soon as the size category it belongs to is determined then an enormous number of species can de discounted and the road to identification becomes much easier. A good system to use is the phenology of the species we wish to identify, and by this we mean the time of year it crosses the Strait in one direction or the other. If we see a raptor in February and we're not sure if it's a Honey Buzzard or a Common Buzzard, then we can assure you there's more than a 99% chance it is a Common Buzzard because Honey Buzzards are not seen in the Strait until April. Another thing to bear in mind is if it is travelling in a flock or alone. An extreme example would be if we see a faraway flock of about a hundred birds, we can be certain they aren't Sparrowhawks or Ospreys. This may seem obvious but, on spotting the flock, the observer may unconsciously eliminate a number of possibilities and concentrate on the ones causing some doubt. The way a bird flies also helps us to identify it because, and resorting to an extreme example again, rapid, continuous wing flapping cannot be that of a Griffon Vulture, which is slow and heavy. But in spite of all the experience you gather over the years, there are still times when migrant birds surprise you, as in the case I mentioned earlier of the 101 Short-toed Eagles flying together in a thermal.

For the benefit of those of you who are just beginning to observe birds in flight, let me describe the most common species of soaring birds crossing the Strait. This doesn't mean that experts can skip this part, because I am also going to mention some aspects of birds' behaviour that are almost unique in bottlenecks such as the Strait. To make identification even easier you can marvel at the spectacular drawings of Enrique Navarro. He is an architect by profession and his amazing drawings of birds in flight are works of art in themselves, giving a whole new dimension to this book.

In the Strait, when we refer to kites we mean the Black Kite, because the Red Kite is just an accidental visitor. The Black Kite is a medium-sized bird with a dark belly and long wings and a lighter coloured head which stands out. The tail in juveniles has lighter coloured undertail coverts and darker tail feathers. The primaries are generally lighter than the secondaries except for the darker "fingertips", although this apparent light patch be absent in many. Its tail is long and narrow and when open has a distinct "kite's tail" tip. Another feature of the tail is that the bird can tilt it according to the direction and strength of the wind and it is the only raptor that does this so obviously, although you may detect this tail movement in the Short-toed Eagle. The juveniles' underparts are generally lighter coloured than adults' and inexperienced ornithologists may mistake them for Red Kites, but their "newer" feathers can be seen clearly. On the plumage on the back of juveniles there is a very definite open M shape. They usually fly in more open flocks than Honey Buzzards but giving a definite impression of cohesion, although when the wind gets up their flapping becomes less vigorous and more clumsy. Seen head-on their wings seem to droop slightly. From a distance when they look very tiny they can be mistaken for Honey Buzzards, but the stronger, deeper wing flapping of the latter should clear up any confusion. A lone Black Kite can also be mistaken for a dark phase Booted Eagle, as we shall see later on.

The Black Kite is one of the commonest and most gregarious of the migrants crossing the Strait.

Egyptian Vulture *Neophron percnopterus*

The Egyptian Vulture, whose numbers are unfortunately on the decline, may be mistaken, by beginners, for a White Stork when seen from afar.

The adult Egyptian Vulture is a large raptor with a notably small, yellowish and featherless head, and a long narrow beak. Underneath, the adult has a white belly and coverts and black trailing edge feathers, like the White Stork or Booted Eagle, but it can't be mistaken for these because its tail has a very clear wedge shape. It has long, wide wings and beats them slowly but constantly, giving the impression of being a heavy, but at the same time, powerful bird. Seen head-on its wings are spread straight or arched and drooping at the tips. It is rarely seen in a flock and, if it is, the flock is very small. Depending on its age it can have one of four different plumages. As a juvenile it is completely blackish brown except for its tail which against the generally dark tone of the bird seems lighter coloured and, sometimes, when visibility is good, has a slightly orange tone to it. As an immature it is generally lighter in colour, especially its back, and has a few white feathers. The subadult is similar to the adult but with some dark patches on its back and the adult has white belly and coverts, black trailing edge feathers with some white patches on its back.

The Griffon Vulture, instantly recognizable by its large size and slow beating of wings.

The Griffon Vulture is almost unmistakeable because of its size and is only similar to the Black Vulture and Rüppell's Vulture. Its wings are long and wide, it beats them in a slow, heavy movement and they are V-shaped when gliding. Its tail is very short and sometimes looks wedge-shaped when rather worn. When they arrive at the coast the flocks are well spread out, but once over terra firma they form compact groups in thermals.

Short-toed Eagle *Circaetus gallicus*

The Short-toed Eagle is known in Spanish as the *culebrera* or snake-eater. It is a large raptor with long, wide, rounded wings. It has a large head and its body is, on the whole, white in colour. Its tail is fairly long and narrow in flight. Arriving at the European coast on prenuptial passage it can often be seen with its bill open, a sure sign of tiredness and stress. Its flight is slow and lazy-looking with deep, powerful wing beats. It can be mistaken for an Osprey, but the latter has narrower, longer wings, or for a light-coloured Common Buzzard, although it is larger and doesn't have the buzzard's carpal patches.

As with the Honey Buzzard and the Booted Eagle we can distinguish three different plumage phases. But before we look at them we must point out a characteristic which differentiates it from many other raptors, namely the band which covers its throat and part of its breast. This "bib" may be wide and conspicuous; also narrow and conspicuous or, in a third morph, it may be almost totally absent, in which case the bird is therefore completely white underneath. Some of these eagles have very contrasted plumage, others less contrasted and still others may be all white, and it all depends on the bird's age. In the final stages of prenuptial migration the

Digital montage of different morphs of Short-toed Eagles.

majority of birds are white or their plumage is not very conspicuous and typical of immatures. During the year of writing, 2006, I have observed this but I haven't taken notes because, as I already mentioned, you can't take notes and photographs at the same time.

The Short-toed Eagle is the largest eagle crossing over the Strait in considerable numbers.

The male Montagu's Harrier is a beautiful bird of prey, very fast and a master at soaring.

The Montagu's Harrier is a medium-small raptor with a thin body, long tail and long, slender wings that seem longer than they really are due to their narrowness. Its wings form a deep V shape in flight when soaring and gliding and it is noted for its characteristic swaying. It doesn't migrate in flocks; you may see at most just a few together but it tends to fly alone. You can distinguish different morphs in the male, female and juvenile.

The male has a grey head, throat and breast while the abdomen is almost white. Viewed both from below and above you can clearly see the six external black primaries (its most notable characteristic is the contrast between these black primaries and the light tones of the body), and the thin, dark grey line on the trailing edge of the wings. The underpart of the tail has faint, darker grey bars while on the upperpart these are well marked. The female is smaller with narrower, darker wings, very similar to the Hen Harrier and juveniles. It has a very light coloured rump, unlike the juvenile, and its underwings are striped with dark edges. Juveniles are very similar to female Hen and Montagu's Harriers but with chestnut coloured underparts and dark

primaries and larger coverts. There is a morph which looks to be black when it is difficult to distinguish between male and female unless at a short distance and in good light. Under such favourable conditions you can make out the distinguishing patterns for both sexes, but it's not easy. On prenuptial migration between the end of March and early April you can see them quite often and also at the end of postnuptial passage.

Marsh Harrier *Circus aeruginosus*

Crossing the Strait is no problem at all for Marsh Harriers.

The Marsh Harrier is a medium-sized raptor, the largest of the harriers, and similar in size to a buzzard but with a narrower head and body. Its wings are long with parallel edges and rounded tips like a buzzard's. Its tail, typical of all harriers, is long, narrow and rounded at the edge. In flight, also a harrier characteristic, its wings form an open V shape and they are raised when soaring. The male's belly is brownish, and looks dark from a distance, and the primary tips are black. On the upperpart of the female the crown, nape and lesser coverts can be yellowish, as can the throat. Juveniles are mainly dark brown. They normally fly alone and when soaring they open their wings more than other harrier species. From a distance they can be mistaken for a Black Kite but you can tell them by their wider, more rounded wings as well as their rounded tail and, when soaring, their wings are an open V shape. Females have a distinctly light coloured head, as is the leading edge of their wings.

The Sparrowhawk is a small bird of prey with short, broad, rounded wings. It has closely barred underparts, giving the impression of black on white, likewise its long tail, except for the underwing coverts which are white with no bars. The male is smaller than the female and has reddish or orange-coloured bars on its flanks. Its back is a sort of dark to bluish slate grey colour. They fly with short, rapid wingbeats alternating with short glides. It wouldn't be unusual to see a Levant Sparrowhawk in the Strait but it would be difficult to spot because ornithologists who come here are not used to it, it is a relatively small bird and they fly at quite a distance. But I'm sure that some year someone will see one and maybe photograph it too.

The Sparrowhawk is a small, rapid-flying raptor which crosses the Strait alone.

Common Buzzard *Buteo buteo*

The Common Buzzard, possibly due to climate change, has all but disappeared as a migrant over the Strait.

The Common Buzzard is now just known as a buzzard, as it has become very scarce of late, as is the case with other buzzard species such as the Long-legged and the Rough-legged Buzzard. It is a medium-sized bird of prey, which, together with the Honey Buzzard has the widest variety of plumage tones, ranging from white to black and every possible tone of grey. In the Strait the most common birds are dark in colour, typical of southern latitudes. The main characteristics of a Common Buzzard are broad, rounded wings, a broad, not very prominent head and a short, rounded tail. On the underparts you can see a dark band on the trailing edge of the wings and a subterminal band on the tail. On its breast you can see a U shape on some birds which is lighter in colour than the other feathers. An in-experienced ornithologist may take this bird to be a Honey Buzzard. The difference lies in the shorter, more robust head and shorter tail of the Common Buzzard and when soaring it lifts its wings, whereas the Honey Buzzard keeps them flat, and its wing beats are shallower too. Some-times it can be confused with a dark phase Boot-ed Eagle, but the eagle has five primaries and the buzzard just four.

The Booted Eagle is the smallest of the eagles in Iberia and has different plumage forms.

The Booted Eagle, known in Spanish as "calzada", is a medium-sized raptor. It has a somewhat smaller head and neck than the Buzzard and parallel wing edges with a slightly curved trailing edge. Seen from below, the most outstanding feature is the contrast between the white undertail coverts, white or lightly speckled body and the almost black trailing edge feathers and the lighter coloured inner primaries, which form a very conspicuous light patch in some birds. Its tail is dark grey with an even darker tip. A juvenile's belly is more of a darkish red than the adult's. Its back wing coverts are much lighter than the trailing edge feathers and form a characteristic band along with the same tone of scapulars. Its rump contrasts with the light tail coverts and the blackish tail. The dark phase Booted Eagle may be confused with a Black Kite, but its tail is rounded, its primaries are blackish and there is a light patch in the inner primaries. There are also two little-known intermediate morphs, details of which can be seen in the photograph and all four morphs can be compared. A flock of Booted Eagles is not compact and not very numerous and they beat their wings continuously. Seen head-on it is the only raptor with two clear white patches on both sides of its neck. It has a habit of ruffling its feathers in flight and when soaring it looks back much more often than the Honey Buzzard does.

Digital montage of different morphs of Booted Eagles.

Osprey *Pandion haliaetus*

The Osprey is a large to medium-sized raptor with long wings and a short tail. Its head is small in comparison to the rest of its body and it has a relatively long neck. When soaring in front of an observer it looks like a seagull with somewhat arched wings. It has a constant wing beat and alternates with long stints of gliding. It may be confused with a Short-toed Eagle, although it is smaller and has longer wings and neck. You can tell an adult from a juvenile by its very conspicuous breast band.

The Osprey, when migrating, does not need to cross via straits.

Phenology, migrating times and local, seasonal meteorology.

In the Strait of Gibraltar you can see birds on migratory passage in either direction every month of the year, but in varying numbers. Phenology is the study of these recurring times of passage, so that we know, fairly precisely, which species of birds are migrating at any given time.

There are two key times of year for visitors wishing to watch migration in the Strait, the prenuptial migration in April and the postnuptial in September. This doesn't mean that they are the only times of year to see migrants but it is when there are most species to be seen together. If we go to the Strait in November we can see vultures migrating but nothing else and if we arrive in the second half of July we'll see a lot of White Storks and also Black Kites, but only these two species. In April, on the other hand, we'll see every species of soaring bird which crosses the Strait and the same again in September.

A winter storm gathering at dusk, threatening to unleash itself on Punta Paloma (Tarifa). In the background Jebel Musa in Morocco.

When deciding to go and watch the migration of soaring birds in the Strait we have to ask ourselves what we want to see, because our preferences may not coincide with those of other ornithologists and we may arrive at the wrong time to see what really interests us. To guide an observer coming to the Strait for the first time we're going to have a look at migrating species month by month, ending with a table as a kind of summary to give an overall view.

Just as the meteorological year begins in September, ours begins on the first of February for one simple reason; the prenuptial migration of the White Stork has finished towards the end of January, although some flocks cross right up to the end of June.

With the odd exception from year to year the first migrant birds to cross the Strait on spring or prenuptial migration *en masse* are Black Kites together with Short-toed Eagles. This migration starts in the second half of February and there is one special aspect of it compared to others. In general, except in the case of Honey Buzzards, migration in either direction starts off with a few individuals, then small flocks before finally getting really underway with compact flocks, large both in size and numbers. The prenuptial migration of both Black Kites and Short-toed Eagles starts, we could say, with a bang and no build-up. So, from hardly seeing a migrant at all (except for White Storks which cross the Strait in either direction almost throughout the year), you suddenly see tremendous flocks of these birds. In February you also start seeing some Black Storks and the odd Egyptian Vulture too.

On days when there's a *poniente* wind Black Kites use a certain strategy to cross the Strait. This wind starts dropping as the evening draws

on and increases in strength during the course of the morning. Black Kites know this and start crossing at dawn before the wind gets up. It is not unusual to see them arriving on the European side at first light and as the morning wears on and the wind strength increases the number of flock arrivals diminishes. This doesn't happen with Short-toed Eagles, which are bigger and heavier and have to wait for thermals to soar and gain height without much effort and then cross. In their case, in February, depending on whether it's a warm day, or cloudy, or there's a strong wind blowing etc., you don't often see these magnificent eagles arriving before 10.00 a.m.

Everyone knows that the weather in February is unstable and is locally known as *febrerillo el loco* (crazy little February). You get very windy days alternating with short periods of rain. Winds are from the west, especially W and SW. Radical changes in wind direction from one day to the other are the norm. The few cold days in the area are already past and when you start to feel the sun on your back you can really spend some enjoyable days watching the migration.

The month of March is one of the most important months for prenuptial migration in the Strait. Days are noticeably longer and after some time in the sun you have to take off your sweater and you need a cap to keep the sweat off your brow. This is the climax in the passage of Short-toed Eagles, Black Kites, Marsh Harriers, Sparrowhawks, Common Buzzards, Ospreys, Egyptian Vultures, Black Storks and, surprise, surprise, the White Stork, which keeps on crossing. Booted Eagles and Montagu's Harriers also start crossing at this time. You have to watch out for the *levante* storm winds which hold up the migrants, especially the less skilled ones such as the Black Kites and Short-toed Eagles. Once

the storm is past and we're lucky enough for the wind to drop, we may witness the amazing passages of these two species, which have been waiting for more favourable weather conditions at the bottleneck in Morocco. A good time to go to the coast to see them arriving is 8:30 a.m.

But if I had to recommend a month to an ornithologist, who is a stranger to the area, then, for many reasons, it would be April. This month is spring in the south of Spain, meaning it's hot and we need a short-sleeved shirt, a bottle of water, a hat and, something I haven't mentioned so far, but is a must to watch migration from: a folding beach chair. Those coming down from the interior, apart from binoculars, cap and canteen, need to bring sun cream otherwise they'll get sunburnt. While Short-toed Eagles, Black Kites, Marsh Harriers, Sparrowhawks, Common Buzzards, Ospreys, Egyptian Vultures and Black and White Storks are still arriving, but in fewer numbers, they are joined by Montagu's Harriers and Griffon Vultures. The passage of Booted Eagles is now at its height. In the second half of the month you start seeing the first Honey Buzzards. One of the most spectacular migrations is that of the Griffon Vulture and you can already see flocks of up to a hundred or so, and several of them, in one single day. The weather is still unsettled, quite normal for the season, and really windy days alternate with very heavy but short-lived showers. We mustn't forget to mention that this is the time when flowers bloom, both on the coast and inland. *Iberis gibraltarica* has already bloomed on the Rock and in the meadows of Los Alcornocales beautiful orchids such as *Ophrys tenthredinifera* and the first rhododendron flowers appear.

In May we come to one of the most important phases of prenuptial migration and that is

Opposite

Photographers in a meadow near Tarifa Maritime Rescue Centre, awaiting the arrival of prenuptial migrants in May 2006.

the passage of Griffon Vultures and Honey Buzzards. The latter because flocks of up to 300 birds are arriving at the coast and the vultures because it's always a major spectacle to behold when these enormous birds arrive fighting against exhaustion and flapping their tired wings, while being chased by seagulls. In the first fortnight you usually get *ponientes* of force 3 to 5, although latterly from 2002 to 2004 we've had a predominance of *levantes*. With these winds you can observe migration at your leisure, from the mouth of the river Guadalmesí to as far as the Rock of Gibraltar. It is the month favoured by tourists from the rest of Europe for several reasons: if the ornithologist is accompanied by a non-ornithologist partner, then the latter can spend time, for example, sunbathing on the beach; usually there are no windy days, the temperature is good and the flowers in bloom are quite spectacular. You can still see Short-toed and Booted Eagles, Black Kites, maybe an Egyptian Vulture, Sparrowhawks and, on rare occasions, Marsh and Montagu's Harriers. The whole coastline from Cabo Gracia right to the Rock is a tapestry of flowers in bloom. The time to arrive at the coast is 7:00 a.m. if we don't want to miss a flock of Honey Buzzards, which, like the Black Kites, cross the Strait early, although the latter don't mind which way the wind is blowing.

All good things come to an end and June sees an end to all the spring colours and, almost, to prenuptial migration. The only birds still crossing are some Griffon Vultures up to the end of the first fortnight and the odd small flock of White Storks. You might still see a Short-toed and Booted Eagle or two, but passage as such has finished. Very windy days, if any, are few and far between and not as fierce as in winter. The heat begins to make itself felt and the fields

are becoming parched, changing from emerald green and multi-coloured May to ochre and brown colours more typical of desert regions. You'd think green had never existed, it was just a dream and now we're in Africa's torrid lands, although it's not the relentless Andalusian heat you get further inland. If we arrive at 8:30 a.m. migrants may still not have passed overhead.

Come July the summer is in full swing and those of us who are not beach fans just have to wait for the second fortnight to rediscover what we like doing most, which is watching and photographing birds. Autumn or postnuptial migration starts, even though autumn sounds a long way off, and now it's time to watch birds leaving instead of arriving, as we have been doing up to now. In early July the skies are only occupied by irritated seagulls, patrolling up and down, parallel to the coast, with no enemies to harass. The second half starts off with some small flocks of storks, comprised of 20 to 50 birds, but already after 5 or 6 days we begin to see flocks of several hundred. Kites appear around the 20th of July and flocks are large and plentiful. Sometimes, towards the end of the month, there's a lull and migration, which started with the arrival of juveniles, comes to a stop and for a few days you hardly see any flocks on the coast at Tarifa. Let me insist again that those non-ornithologists can have a really good time on the beach while their partners are busy with migration! The *levante* wind may blow strongly sometimes, but it doesn't normally last for many days and what you really notice is the heat. To see the first flocks of White Storks you don't need to get up early – 9:00 a.m. is early enough to be in Tarifa.

August, the main holiday month in Spain, is also the main time for the passage of White Storks and Black Kites. Towards the end of

Dwarf Morning Glory (*Convolvulus tricolour*) in flower
in a meadow at Jimena de la Frontera in May 2006.

Greater Flamingo chicks at the lagoon in Fuente de Piedra in July. At this time young migrants of other species are already crossing the Strait.

A flock made up of juvenile storks flies nervously over Los Lances beach near Tarifa.

the month the first Booted Eagles, Short-toed Eagles, and a few Ospreys appear. Montagu's Harrier is seen in abundance now and you also see the first Egyptian Vultures and Honey Buzzards. The fields look as if they've been burnt, they're absolutely bone dry and the snails "flee" from the hot ground and climb up anything they find looking for moisture. There may be a day or two when the wind puts our beach chair's stability to the test, but not many. At 8:00 a.m. at the observatories you can be sure to meet up with ornithologists who've come from all over, anxious to watch this great spectacle, and you won't be let down.

Autumn migration reaches its climax in September both in the number and variety of species to be observed, and in the increasing number of ornithologists who come down to the Strait. On Saturdays and Sundays bus-loads of them arrive, armed with their optical instruments and the latest books on birds. It's the month when ornithologists from Britain, Holland, Germany, France, Portugal etc. expect to see those species that can only occasionally be seen elsewhere. The passage of Black Storks, Booted and Short-toed Eagles, Honey Buzzards, Ospreys, Egyptian Vultures, Sparrowhawks and Montagu's Harriers is at its peak and, at the beginning of the month you can still see a flock or two of White Storks. The curtain comes up and the show starts at 8:00 a.m. with a large number

The Strait of Gibraltar in spring 2006. In the foreground the Natural Park of the Strait and in the background the coast of the Kingdom of Morocco.

and variety of species flying over our heads. If we keep our eyes peeled we may see an immature Spanish Imperial Eagle, a Golden Eagle, Spotted Eagle or Lesser Spotted Eagle and other rarities which cause so much excitement among many ornithologists. There is a radical change in the weather during this month and many days when big, black clouds appear to cool down the atmosphere, so you need a sweater on early in the morning until the sun comes out. However, it doesn't often rain, at least to any degree and winds are changeable but with a predominance of *levantes*, which are typical for August, September and October.

October is the month when ornithologists desert the area. You can still see a few Black Storks, Booted and Short-toed Eagles, Sparrowhawks … but there's not much point in spending much time watching out for them because the real show is over. Nevertheless, when something ends something else may begin and that's what happens with the migration of Griffon Vultures, Common Buzzards and Marsh Harriers, which choose this month to cross the Strait. Marsh Harrier migration isn't very spectacular but it is interesting to see them around the former La Janda lagoon and, from the beginning of the month you start to see unusual numbers of Griffon Vultures together. At the end of the month the first flocks of these huge birds start crossing, but I think this particular migration is of interest only to real diehards because, as I've mentioned before, several days may go by without one flock crossing. You can still see the odd flock of White Storks going over and, paradoxically, the first flocks start coming back from the other side of the Strait. One 4th October I made a note in my field diary of a flock coming in over Tarifa, and several others in the days that followed. Do

incoming flocks cross outgoing flocks on their way over the Strait? I wonder. It's now typical autumn weather and the first rains appear. Near many observatories you can now see the strange flower of the mandrake (*Mandragora autumnalis*). In theory the migration of the Common Buzzard is now at its peak, but you no longer see those flocks that Bernis talked about in the seventies. They say it's to do with climate change. I haven't seen a flock of ten together for years. In mid-October the first autumn rains cool down the atmosphere and the scenery begins to take on that autumn look: white clouds scattered all over the sky, giving depth and volume to the autumn skies.

In November you can only see vultures crossing over, slowly, but surely, to the Moroccan coast. You also start to see flocks of White Storks arriving in greater numbers on the European side of the Strait. At each side of the river Palmones where the A-381 road joins the N-340 there are flocks of White Storks next to a small, rainwater-filled pool. Officially this is the rainiest month in the area, with over 30% of the total annual rainfall and windy, rainy weather is quite common.

December is White Stork migration time, especially in the second half, peaking at the end of the month. Around Christmas temperatures are at their lowest, although rarely do they drop to freezing point and frost is even rarer still. Windy, rainy weather is the order of the day but when the sun appears from behind the clouds you don't need winter clothing, which is quite light in the area anyway.

January sees the end of the migratory year and it is when you see most migrant White Storks at the Miramundo and Los Barrios rubbish dumps, although they still keep on arriving

A group of Glossy Ibis resting in the rice fields in La Janda after completing their migratory flight.

for several more months. At the end of January the first Lesser Kestrels take you by surprise because you don't see them cross the Strait. Squally weather with heavy rain and strong wind is the norm.

Observatories and recommendations for ornithologists

There are two types of observatory in the Strait: those built by the Andalusian Regional Government, vandal- and earthquake-proof, and those that ornithologists prefer and which haven't yet been built for several reasons, which we will outline later. However, it's no good knowing where the observatories are if we don't know which one to pick and going from one to the other until you find a migration flyway is no way to carry on. Besides, you have to choose an observatory according to whether it's prenuptial or postnuptial migration you want to observe and it also depends on what you actually intend to do.

To begin with I would recommend to any ornithologist arriving in the area for the first time to contact one of the local ornithologist associations because they are people with a lot of experience in migration and can put you on the right track. Alternatively, if you're unable to make contact, then call Salvamento Marítimo(Sea Rescue Service) at 956681001. They give local shipping the weather forecast, especially as regards to the wind strength and direction. It's a very reliable forecast and is given by hour, so if you call at 8:00 a.m. you'll know what the wind is and what it will be during the next few hours. The worst time for migration is when there are changes in the wind, which can be quite often, because you don't know how the migrant birds are going to respond. Sometimes this phone number is engaged for so long that I prefer to go outside and find a place where I can tell clearly where the wind is blowing from. If we're in Algeciras, it's enough to look towards the area of the Gibraltar oil refinery and see the wind direction from the smoke coming from the chimneys. If, from our position, the smoke is to the right of the chimneys, then the wind is basically *poniente* (W and NW) and if it's to the left then it's *levante* (E and NE). Seen from La Linea or Gibraltar then it's the other way round. There's no such reference point in Tarifa but you can tell from the movement of the many flags to be seen between Tarifa and the Valdevaqueros inlet. If you're going towards the inlet from Tarifa and the flags are facing to the left of the flagstaffs then a *levante* is blowing, and if to the right, then it's a *poniente*. We're so used to the winds in the area that little details that might go unnoticed to a visitor are quite meaningful to us. For example, if I'm going into the underground garage where I, and many people, park our cars and the door tends to close again by itself then the wind is a *levante*, but if it stays open and takes an effort to close it, then it's a *poniente*. If at home the curtains blow towards the inside then it's a *poniente* and if they blow out through the window then it's a *levante*. There are even streets in the town where, depending on the direction of the wind you know if it's *levante* or *poniente*. So, the main thing is to get to know the wind strength and direction to decide the ideal place to be.

On postnuptial migration you find a lot of ornithologists to consult both in the countryside and at the observatories, and if you're at the wrong observatory there will be people coming from others with information about where migration is more interesting. The arrival of the mobile phone has also contributed to an exchange of information between observatories so that at any given time you know what species are passing over which observatory.

On prenuptial migration it's quite normal not to meet anyone at all anywhere on the coast. The first ornithologists I see are in April and usually from northern Europe, i.e. Holland, Germany and Great Britain. I have met Dutchmen and Englishmen who have spent the whole of May at a camp site in the area and haven't missed a day's passage. Finding the right observatory is more difficult because we have to rely on our own information, but, as we shall see, it's really gratifying to find things on your own. To do so, you need a good map of the area and, before trying to watch birds, you need to spend a day seeking out the most important places, whether they are official observatories or not.

The difference between observing prenuptial and postnuptial migration is the positioning of the observer in relation to the coastline. In spring you have to get as near to the sea as possible to see the birds arrive close to, and to do that you have to locate places that are nearest the shoreline and, most importantly, as high as possible above it. That's why we never choose those concrete-built observatories because most of them are inland or, like the one at Guadalmesí, not high enough above the sea. You must also

Ornithologists, both Spanish and from abroad, at the Cazalla observatory in Tarifa.

bear in mind that the Natural Park of the Strait is still within a military zone and there are still areas with no access because of gun emplacements. One of these is called La Hoya at the Bujeo pass, from which you have a great view in both directions and is therefore a great spot for both spring and autumn migrations. Unfortunately you're not allowed to go there. La Hoya is also the name of the place where the only wind farm in the park is situated. It belongs to the same military zone but is at a much lower altitude.

In spring when a westerly (*poniente*) wind is blowing we position ourselves between the mouth of the river Guadalmesí and Gibraltar, as near to the coast as possible. As the wind gets stronger we go on to Gibraltar, where, with a force 4 to 6 wind on the Beaufort scale you can watch migration at the lighthouse at Europa Point. If the wind is a *levante*, from the opposite direction, then we go from Guadalmesí to Sierra Plata, this being a long continuous stretch of land, whereas in the other direction you have the Bay of Gibraltar. When this wind is force 6 or 7 a good place to watch migration is from Ombligo, a viewpoint between the Torre de la Peña camp site and the Valdevaqueros inlet, next to a small pine wood, as far as the Ranchiles rock in Sierra Plata above San Bartolomé. You sometimes, but not often, get north winds in spring and north-easterlies when migration tends to be concentrated in the Tolmo-Faro area and then a good place to watch from is between Punta Secreta and the Guadalmesí river mouth.

Whenever anyone asks me, I always say I prefer prenuptial migration to postnuptial, without wishing to detract from the latter at all, because it gives me the chance to see the migrants from above, except perhaps in the case of White and Black Storks and Honey Buzzards, and, although I don't consider myself to be antisocial, I'm not keen on observatories when they're full of ornithologists because I just can't concentrate or take my mind off what's going on around me, and my work suffers. There's also more action in prenuptial migration and a photographer appreciates things like attacks by seagulls and kestrels, with vultures and Short-toed Eagles falling into the sea. Then there are those unusual shots of migrant birds such as raptors against a sea background or passing ships, flying above built-up areas or below the photographer, many of which you can find in this book.

Postnuptial migration is easier to pinpoint because the observatories are further inland and, as I said before, there are lots of ornithologists to ask. If you drive from the Algeciras area towards Tarifa, the option I would choose is the Algarrobo observatory at Huerta Serafín, about 100m past the km 99 marker on this road. According to local legend, Serafín was a retired pirate who bought this farm and buried his treasure there. Anyway, from the "official" observatory you can see the chimneys of the petro-chemical works in the Bay of Gibraltar and you will remember that smoke blowing to the right means *poniente* and to the left *levante* wind. I must remind you that most of the Honey Buzzards and Black Storks crossing the Strait pass over this observatory and it's a great place to watch birds when there's a *poniente* wind. If this wind is blowing and you have an off-road vehicle then there's another highly recommendable observatory quite near Algarrobo, but a bit difficult to get to. Driving from Algeciras to Tarifa just after Huerta Serafín (Algarrobo) there's a track on the right and if you drive up it you come to a promontory from

Opposite

Photographers next to the Tarifa observatory, taking shots of Honey Buzzards arriving in spring 2006.

where you can see Huerta Serafín far below. The advantages are twofold: it is much higher than Algarrobo and, especially at weekends when there are lots of ornithologists about, you won't find anyone at this spot.

A new development, with the inevitable golf course to boot, is now threatening to invade the privacy of birdwatchers when they find themselves surrounded by buildings, shopping centres, schools etc.

I mustn't forget to add that the little carob tree (*Ceratonia silicua*), next to the observatory, was planted thanks to Cristina Parkes, whose strong character and steadfastness were too much for the cows and goats that sought shade under it during the hot months. Time and again she went there to water and care for it, so that it wouldn't be the main course for those animals! By putting a mesh wire fence round it we now have a carob tree at the observatory of the same name (Algarrobo). What a lesson in tenacity!

Another fine observatory which is also very popular with ornithologists is the one at Cazalla, although it is not officially recognized as one. I'll explain: when the Andalusian Government was considering where to build observatories they sent a technician to pick out the most suitable spots and among them was, of course, the hill at Cazalla. When he approached the local councillor in Tarifa to arrange the lease of the land, the councillor refused to have an observatory built there. So what we have now is one of the most important places for migration in the western Palaearctic with no observatory because a local councillor, who obviously hadn't a clue what he was being offered, refused to have one.

As you drive from Algeciras to Tarifa you go over an S-shaped bridge and up a very steep hill and on the right on a mound there's a building made of concrete; this is Cazalla hill, just a few metres before km 87. There you leave the main road, drive up a short tarmacked slope and you come to a flat piece of land where you can leave the car and enjoy the view. This is the ideal observatory for days when there's a *levante* and you can watch the passage of White Storks, although just a bit away from the flyway they usually use, but good enough to watch them approach their crossing point. The building storing butane gas bottles limits all-round 360º vision. But there is a little building, that Tarifa Town Council has lent to COCN (the Black Stork Ornithological Association), along with a couple of garden benches, which is used as a base by ornithologists from the Migres Programme. It is without doubt the best observatory in the Palaearctic, which thousands of soaring birds pass over on both migrations and, together with the ones at Algarrobo and Huerta de Serafín, is the one most frequented by ornithologists who come to the area for the postnuptial migration.

As I write this book Tarifa Town Council has announced that it wants to build an observatory and a Migration Interpretation Centre plus a restaurant and cafeteria in a building with a viewpoint on its flat roof. I hope it is named after Prof. Francisco Bernis who did so much to draw attention to migration in the Strait when nobody cared much about it but received very little official recognition for his efforts. I believe it's the least they could do in his memory.

As well as these two observatories for postnuptial migration there are others nearer the coast, where vision is greater because there are no nearby promontories to block it, but as they are at a much lower height then the birds fly high above you. A timely reminder to readers: the strategy of soaring birds is to gain height at the shoreline they leave from to try and reach the opposite shore by gliding as much as possible and thus save energy. So, on postnuptial migration, they leave the European side flying high or very high, except on extremely windy days, when they have to come down, because the lower they fly the less wind strength they have to face. I must also point out that each activity must be considered separately: a magnificent observatory for bird watching may be a poor one for photographing them and it may also be an ideal place to count migrant birds but not to observe them for the sheer pleasure of doing so. As an example let me say that there's an observatory – but no platform built – at the Cabrito bunker and it's wonderful to watch prenuptial migration from there but no good to take photos from.

The observatory known as Trafico, refers to the area where it was built, near the Tarifa Trafico (shipping control) building, just outside Tarifa. It is an ideal spot on windy days because migrants follow the coastline and you can watch them quite comfortably as it stands on a promontory. It's a great place to observe White Storks as the majority of them pass close by.

How is the observation of migrant birds organized in other countries fortunate enough to possess this wonderful resource? Perhaps the most conspicuous and famous example in the world is the Hawk Mountain Sanctuary in the United States. Its director is now Dr Keith L. Bildstein, who was kind enough to write the prologue to the English edition and whose experiences he now describes. He visited the Strait a few years ago and was amazed by what he saw, so much so that he comes back every year.

Hawk Mountain Sanctuary

Keith L. Bildstein.
Sarkis Acopian Director of Conservation Science,

Hawk Mountain Sanctuary, the world's first refuge for birds of prey, was founded in 1934 by conservationist Rosalie Edge who wanted to stop the shooting of thousands of raptors migrating along the Kittatinny Ridge in the central Appalachian Mountains of eastern Pennsylvania, 170 km west of New York City.

Today, Hawk Mountain Sanctuary maintains the longest and most complete record of raptor migration in the world. In addition to its colorful past, the Sanctuary offers magnificent views of 16 species of migrants, including regionally scarce and seldom-seen species like North Goshawks and Golden Eagles, as well as panoramic views of Pennsylvania's central Appalachian countryside. Modern ancient mountains and modern weather patterns are responsible for the large numbers of migrants that pass Hawk Mountain's lookouts each autumn.

The region's mountains rarely extend more than 300 meters, above the surrounding landscape, but what the ridges of the more than 300-million year-old central Appalachians lack in height, they more than make up for in length. The Kittatinny Ridge, which includes the promontory called Hawk Mountain, alone extends across 400 kilometers northeast-to-southwest from southern New York State to southeastern Pennsylvania. The entire Appalachian Mountain range spans 20º of latitude, or almost one-fourth the distance from the Equator to the North Pole. The range's exceptional length, together with its largely north-south orientation, creates the backdrop for one of the raptor world's great migration flight lines. Hundreds of thousands of raptors use this important updraft corridor each autumn. Northern breeders include Peregrine Falcons from Greenland and Labrador, and Ospreys, Sharp-shinned Hawks, and Broad-winged Hawks from Eastern Quebec. Southern breeders include Turkey Vultures, Cooper's Hawks, and Red-shouldered Hawks from New England and the Mid-Atlantic States.

Raptor migration at Hawk Mountain is most pronounced when northwesterly winds strike the northeast-to-southwest oriented Kittatinny at right angles. Deflected up and over the ridges, the updrafts form an aerial highway for soaring birds of prey that creates in a leading line for thousands of outbound migrants each autumn.

The Sanctuary itself owes its origins to the irruptive movements of Northern Goshawks along the Kittatinny during the winters of 1926-27 and 1927-28. The goshawk was considered a vicious killer at the time, even by many conservationists, and its "invasions" were not welcome events. Rural residents reported the unappreciated irruptions to local game protectors, and they, in turn, alerted the State Ornithologist, who visited Hawk Mountain in October 1927. The ornithologist and his companions collected several shot raptors, including four goshawks. Accompanied by several shooters, three days later the group secured "in a remarkably short time" 90 Sharp-shinned Hawks, 11 Cooper's Hawks, 16 Northern Goshawks, 32 Red-tailed Hawks, and 2 Peregrine Falcons. No wonder the local residents had named the site "Hawk Mountain."

The State Ornithologist published his findings in a technical ornithological journal the following June. The 12-page report had two diametric effects. The first was that the Penn-

sylvania Game Commission used the report to justify establishing a $5 bounty on Northern Goshawks. The second was to alert ornithologists and conservationists to shooting at the site, which set in motion a series of events that resulted in the founding of Hawk Mountain Sanctuary.

Shortly after the article appeared, two Philadelphia birdwatchers, Richard Pough and Henry Collins, visited the ridge-top shooting gallery and confirmed and expanded the initial report in accounts published in several conservation magazines. In the autumn of 1933, Pough displayed photographs of the shooting at Hawk Mountain at an ornithological meeting in New York City. Devout conservationist Rosalie Edge was in the audience the evening Pough spoke. Outraged by what she saw, Edge sprang into action. Together with her son Peter, she toured Hawk Mountain in June of 1934 and, shortly thereafter, leased and then purchased the 565 hectares that was to become the core of Hawk Mountain Sanctuary. That August, Edge hired a young Massachusetts birdwatcher, Maurice Broun, as "ornithologist-in-charge" of the new refuge.

Broun arrived at the shambles of a Sanctuary in early September and spent most of that month posting "no-hunting" signs and informing the neighbors of the site's new status as the world's first refuge for birds of prey. On 30 September 1934 he began counting migrants from what was then called Observation Rocks, and what is now known as The North Lookout. Although Broun initiated the counts principally to document the numbers of raptors that were being protected at the site, he quickly realized that a series of season-long counts would allow the Sanctuary to monitor regional populations

Ornithologists at Hawk Mountain Sanctuary in Pennsylvania (USA) watching autumn migration.

of the birds. By the second year of operation, autumn counts of migrants became the foremost feature of Hawk Mountain Sanctuary's field work.

Broun's reports of substantial flights of Golden Eagles at the site (39 birds in his first autumn), along with the abrupt passage of Broad-winged Hawks in mid- to late September, and the relationship between cold fronts and the magnitude of visible migration at the site, rank among the Sanctuary's most significant contributions to ornithology. By the middle of the 20th Century, Hawk Mountain's long-term counts were being used by numerous conservationists, including Rachel Carson, to document

Pesticide Era population declines in several species of raptors, including Bald Eagles and Peregrine Falcons. And by the end of the 20th Century, Hawk Mountain had established itself as a global information hub and training center for raptor conservationists with its successful international internship program, its authorship of Raptor Watch: a global directory of raptor migration sites, co-published with BirdLife International, and the construction of a world-class biological field station, The Acopian Center for Conservation Learning, for hosting visiting scientists.

Each autumn, counters at Hawk Mountain record a visible passage of approximately 18,000 migrating birds of prey. Three species, Sharp-shinned Hawk, Broad-winged Hawk, and Red-tailed Hawk, represent more than 95% of the flight. The elliptical migrations of many migrants at the site result in a spring flight that is less than 10% of the autumn flight. Although some of the raptors migrating at Hawk Mountain appear as "pepper specks" hundreds of meters above the Sanctuary's 11 available lookouts, many are seen at close range as they slope soar within a few meters of the tree-tops of the Kittatinny's deciduous-coniferous forests. The passage rates of most species increase substantially when and shortly after cold fronts pass through the region and many visitors time their trips to take advantage of this phenomena.

The site's official long-term count is made at the North Lookout, a quarter-acre, boulder-strewn promontory, 1.5 kilometers from the Sanctuary's entrance gate. Activities at the Sanctuary's lookouts include interpretation of the flight, scheduled and impromptu talks on raptor identification and cultural and natural history, as well as ongoing question-and-answer

sessions on raptor migration and conservation. The South Lookout, which is 300 meters from the main parking lots and is wheel-chair accessible. Altogether, 11 Sanctuary lookouts are open to the public.

Although the official autumn count begins on 15 August and ends on 15 December, a few migrants, including Southern Bald Eagles and American Kestrels, can be seen as early as late July, and in most years weather-related pulses of several species, including Northern Bald Eagles, Northern Harriers, and Red-tailed Hawks occur well into December and January. The bulk of autumn migration passes between mid-September and mid-October. The largest outbound movement typically happens during the third week of September at the peak of the Broad-winged Hawk flight.

In spring the official spring count begins 1 April and continues through 15 May. In most years the return flight peaks during the third week of April.

Hawk Mountain's Visitor Center houses raptor exhibits including a gallery with life-size, in-flight models of all 16 of the Sanctuary's regular migrants, historical displays, a bookstore and gift shop, and restrooms. Hawk Mountain Sanctuary maintains a large and routinely updated web site (www.hawkmountain.org) that includes directions to the site and information about restaurants and accommodations in the area, as well as news of the flight and other activities.

Three books have been written about Hawk Mountain Sanctuary: The mountain and the migration by Jim Brett (1991) is a useful guide to the site and wonderful introduction to hawk-watching; Hawks Aloft by Maurice Broun (1949) is a captivating account of the Sanctuary's early years; The view from Hawk Mountain by Michael Harwood (1973) updates the history of the sanctuary through the early 1970s and provides recollections of both hawkwatching and hawkwatchers there. Racing with the sun: the forced migration of the Broad-winged Hawk (Bildstein 1999) describes the flight of Broad-winged Hawks at Hawk Mountain, and Mountaintop science: the history of conservation ornithology at Hawk Mountain Sanctuary (Bildstein and Compton 2000) summarizes the watchsite's contributions to science and conservation.

Nature reserves and areas of ornithological interest near the Campo de Gibraltar

It would be unheard of for an area with such a wealth of biodiversity, scenery, climate, culture etc. not to have other areas of interest in the vicinity. When I refer to this area or "Comarca" I mean the Comarca del Campo de Gibraltar which is a political and administrative division including the municipal townships of Algeciras, Castellar de la Frontera, Jimena de la Frontera, la Línea de la Concepción, Los Barrios, San Roque and Tarifa. Gibraltar, as everyone knows, is not on Spanish territory, but fortunately the ties between their authorities and those on the Spanish side are increasing all the time, as are contacts between Gibraltarians and local Spaniards.

At the mouth of the river Palmones in the Bay of Gibraltar there is an area of wetlands called the Marisma del Río Palmones Nature Area and it belongs to the municipalities of Algeciras and Los Barrios. These wetlands are very interesting as they have various habitats within an area of considerable influence in the Strait, covering a relatively small surface

area of 60 hectares. It contains the last existing sand dunes in the Bay, orchards, pasturelands, wetlands and estuaries with extensive mudflats which appear and disappear with the tides. These mudflats are split by channels with hardly any vegetation, only halophytes and strips of sea rushes (*Juncus maritimus*). There is not much tide variation, being in the Mediterranean, and that explains the type of vegetation in the area. The wide expanse of mudflats attracts a great number of shorebirds and waders when on migratory passage. You often see Greater Flamingos, Spoonbills and the odd Osprey. In fact the only sighting of a juvenile Osprey in the area was here in these wetlands.

The Estuario del Río Guadiaro Nature Area covers 27 hectares and is situated in the municipality of San Roque within the resort of Sotogrande. It is of great ornithological value due to its proximity to the Strait and its relative seclusion. It contains two channels and a sandbar at the beach. It is typically Mediterranean in that there is little tidal variation and because of that there are no halophytes in the vegetation which mainly consists of rushes (*Scirpus maritima*), sea rushes (*Juncus sp.*) and salt cedar (*Tamarix sp.*) etc. It is an area of passage for shorebirds and many species of Passerines and you can easily spot Purple Gallinule, Little Bittern, Coot, etc.

The Breña and Marisma de Barbate Natural Park, covering 3,800 hectares, has three biotopes: a wetland, a pine forest and a coastal strip. The pine forest, containing umbrella pines, was planted in the 19th century to stop the advance of the sand dunes, but the native scrubland has resisted the pressure and typical Mediterranean species, some of them scarce,

are still increasing their presence, such as prickly juniper (*Juniperus oxycedrus*) and Phoenician juniper (*Juniperus phoenicea*). The cliffs or *tajos* rise 100 metres above the sea, the highest being the area called Torre del Tajo which was restored in the 90's. There used to be a unique maritime colony of Cattle Egrets and Little Egrets on this cliff up to the end of the 90's. Records of this colony have existed since the 50's and there were about 1,500 breeding pairs together with Yellow-legged Gulls and Jackdaws which preyed on egret eggs and chicks. It is not known why they stopped nesting there after so many years of successful breeding. However, a new colony established itself in 2004, not too far from the latter one, near the A-48 (Algeciras-Cádiz) and the Vejer de la Frontera crossroads, next to some restaurants and a bodega. Pallid Swifts nest on the upper part of the Tajo and in 1972 the French ornithologist J.M.Thiollay showed me a nest of Peregrine Falcons and another of Tawny Owls. At the beginning of the 90's a pair of Short-toed Eagles settled in the pine forest, but at the moment it is not known whether they are still breeding there.

The marshes are around the mouth of the river Barbate and are mainly notable for many endemic plants growing there. Bird watching is not easy because access is very difficult and the times I have been there I have not been at all satisfied.

Nowadays communication by road and sea makes it easier to get to other interesting places in a matter of hours, so I'll briefly describe some of them so that readers who are interested can make their own enquiries.

Just half an hour away from Tarifa is the Moroccan city of Tangier, the starting point for those who want to visit wetlands in Morocco. If you use Larache as a base then you can visit the Merga Serga Reserve, a really important wetland which acts as a bridge between the Doñana National Park in Spain and the Atlantic coastal wetlands which go down as far as Mauritania and beyond to Senegal.

Less than an hour's drive from the Comarca near Puerto Real (Cádiz) you can find the Complejo Endorreico de las Lagunas de Puerto Real a reserve containing lagoons, one of which, the Comisario lagoon, we have already mentioned. In this area of the province of Cádiz you can also visit the Laguna de Medina Nature Reserve, the Laguna de las Canteras y el Tejón Nature Reserve and the Complejo Endorreico de Chiclana Nature Reserve.

In the north of the province of Cádiz, in the heart of the Sierra de Cádiz, you can visit the Nature Park called Parque Natural Sierra y Pinar de Grazalema, which covers 51,695 hectares and where the Spanish fir (*Abies pinsapo*) is a botanical jewel found only here, in the Parque Natural Sierra de las Nieves, just 10 km from the beautiful town of Ronda (Málaga) and in Sierra Bermeja in Estepona (Málaga).

A little further afield, but highly recommended, is the Parque Natural de la Laguna de Fuente de Piedra (Málaga), a lagoon where, in some years you can see up to 15,000 Greater Flamingos breeding, as well as Cranes and many other wintering species.

The Torcal de Antequera (Málaga) covering 1,171 hectares, is a mountainous, limestone area, part of the Sierra Bética. Water erosion has produced some fascinating shapes in the rocks which lend an air of mystery to the scenery as a whole. There is some spectacular vegetation, many endemic plants and very interesting fauna,

Opposite

Cliff (Tajo) at Barbate. Down below in among the vegetation you can see the colony of Cattle Egrets, Little Egrets and Seagulls which was later abandoned.

Spectacular aerial view of the Strait of Gibraltar.

including an increasing number of south-eastern Spanish ibex (*Capra pyrenaica hispanica*).

The Parque Natural de las Marismas de Cádiz is also a magnificent spot where you can observe all kinds of shorebirds and seabirds, both resting on their migration journeys and breeding.

If a visitor to the area has enough time to go to any of these reserves just outside the Campo de Gibraltar, then it's well worth looking for further detailed information about them.

The Campo de Gibraltar: an area full of surprises

Earlier we mentioned that the Campo de Gibraltar area is comprised of the municipalities of Algeciras, Castellar de la Frontera, Jimena de la Frontera, La Línea de la Concepción, Los Barrios, San Roque and Tarifa, plus the city of Gibraltar. Geographically it's the southernmost part of Andalusia and is the meeting point of two continents, two seas, two cultures, two religions and two worlds.

Communications

Without a doubt, the economic and social development of the Campo de Gibraltar has historically been held back by its poor network of communications. Anyone who looks at a map will see that there is a railway linking it to the rest of Andalusia and Madrid, a dual carriageway connecting it with Eastern Andalusia, a road connection to Western Andalusia –the N-340 - and the coast of Cádiz province and, finally, a recently completed dual carriageway, the A-381, which runs through the Natural Park of Los Alcornocales, narrowing the distance between the area and Sevilla, the administrative capital of Andalusia. There are also a number of ships and catamarans which link the ports of Algeciras and Tarifa with the autonomous city of Ceuta and the kingdom of Morocco. At first sight the Campo de Gibraltar seems to be sufficiently well communicated with the rest of the country and the north of Morocco but the system still has many shortcomings that will have

Plaza Alta in Algeciras. In the background Gibraltar and La Línea de la Concepción.

to be addressed so that the area can be properly integrated into the national rail and road network, which is absolutely essential if full industrial, social and tourist development is to be achieved. These shortcomings are obsolete rail routes, a toll motorway linking the area with Eastern Andalusia and another with Sevilla, a dual carriageway with much too many roundabouts slowing down traffic to the Costa del Sol and an outdated, dangerous road, the N-340, linking it to Western Andalusia.

Now that the A-381 dual carriageway has been finished you can get to this area from Jerez de la Frontera in just over an hour, so Jerez airport is now quite near, as is Málaga airport on the motorway. For British naturalists and ornithologists Gibraltar airport is a perfect gateway to the area. A new arrangement now allows Iberia airlines to use the airport and other airlines will surely follow suit meaning more ornithologists and naturalists will be attracted to the area.

Cities and towns of the Campo de Gibraltar

The city of Algeciras is the Arabic "Green Island" (Al-Yacirat-Aljadra) which King Alfonso XI took from the Moors in 1342 in a crusade-like war in which other Europeans took part. It was reconquered by Mohammed V, King of Granada, in 1367, and razed to the ground. In 1704, after the combined Anglo-Dutch fleet conquered Gibraltar, Algeciras was occupied by exiled Gibraltarians and slowly began to recover. The port is the economic driving force in the town because of its strategic position and deep harbour, which can take large-tonnage vessels. It is one of the main Mediterranean ports for transporting both goods and passengers. From there modern boats and catamarans cross the Strait in times varying from 25 minutes to a couple of hours.

It is a modern city with a population of 120,000, which looks onto the Bay and is built partially on the eastern slope of the hills of Los Alcornocales. It was settled by Phoenicians, Romans, Byzantines, and Berbers, Almoravid and Almohad Arabs and in the city museum you can find many examples of their presence.

The former Nazari fortress belonging to the kingdom of Granada was the site of the original town of Castellar. It was built inside the fortress, well protected by its walls. In 1973 most of the inhabitants left the town to live in the newly built town of Castellar de la Frontera, a typical colonial town situated on a plain at the foot of the old fortress, with much better communications and modern facilities.

The inner part of the old fortress is being reformed and converted into rural tourism cottages. From there you have a spectacular view over the countryside, from the Rock of Gibraltar to the African coast and including some of the White Towns of Cadiz as well as a mosaic of colourful meadows in spring.

The La Almoraima estate, in the municipality of Castellar, is, with its 16,000 hectares, one of the biggest in Western Europe. The former early XVII century Barefoot Mercenary convent building has now been converted into the Casa Convento, a very select hotel with just a few rooms, situated in a perfect spot for the naturalist, writer, business executive or anyone who is looking for the ideal place for peace and quiet.

The Arabs called it Ximena and the Christians, after conquering it in 1431, added "de la frontera" because, for two centuries, it separated the kingdoms of Christians and Nazaris, just like Castellar. Jimena de la Frontera has a castle built first as a fortification by the Romans for the Roman city of Oba. From the VIII century, under Arab sovereignty, it became a truly magnificent castle.

One of the local cave shelters, called Laja Alta, contains some fine examples of Southern Art, mentioned in Chapter II, and you can see drawings of some boats, the only Bronze Age painting with a maritime theme to be found in the area.

The greater part of the municipality is within the limits of the Natural Park of Los Alcornocales, and it also contains the river Hozgarganta, which is perhaps one of the least contaminated in the Peninsula from where it rises until it reaches the town.

For several years now, in the month of November, a congress on mycology has been held in the town and it attracts more and more people from all over Spain. There is also a very high quality Music Festival in summer, in which both virtuosos and amateur musicians get together to perform.

Rock paintings in La Laja Alta on the Altabacar estate in Jimena de la Frontera.

King Felipe V had a series of fortifications built in line, not far from Gibraltar and these have now become the present-day town of La Línea de la Concepción, the youngest town in the Campo de Gibraltar. The only fortification remaining is the Castle of Santa Barbara, which was destroyed by the English when they had taken over Gibraltar, to avoid being besieged by the French in the War of Independence. During the Second World War a series of bunkers were built outside the town, but, as this has grown so much, some of them are now right in the middle!

In this town I've sometimes observed postnuptial migration and been surprised to see flocks of Honey Buzzards flying low over the buildings in the centre. You can also hire boats here to go dolphin watching in the Bay. For sports fans and holidaymakers there are also golf courses and plenty of fairly deserted beaches, as well as spectacular views from the shore of Gibraltar and the Strait.

On the banks of the river Palmones or Cañas you will find the town of Los Barrios, the greater part of whose municipality is within Los Alcornocales. One particular cork oak forest and another, especially, of Mirbeck's oaks called San Carlos del Tiradero, stand out as real botanical and scenic jewels in the south of Europe.

The collection of Neolithic cave shelters in the sandstone rocks has one of the main examples of Southern Art in the area and the Cueva de Bacinete has an impressive number of prehistoric parietal paintings.

From the botanical point of view this town is very important because in its nearby *sierras* our late friend and botanist Betty Molesworth discovered the ferns *Cristella dentata* and *Psilotum nudum*.

Portuguese Oak, whose moss-covered trunk
reminds one of rain forest vegetation.

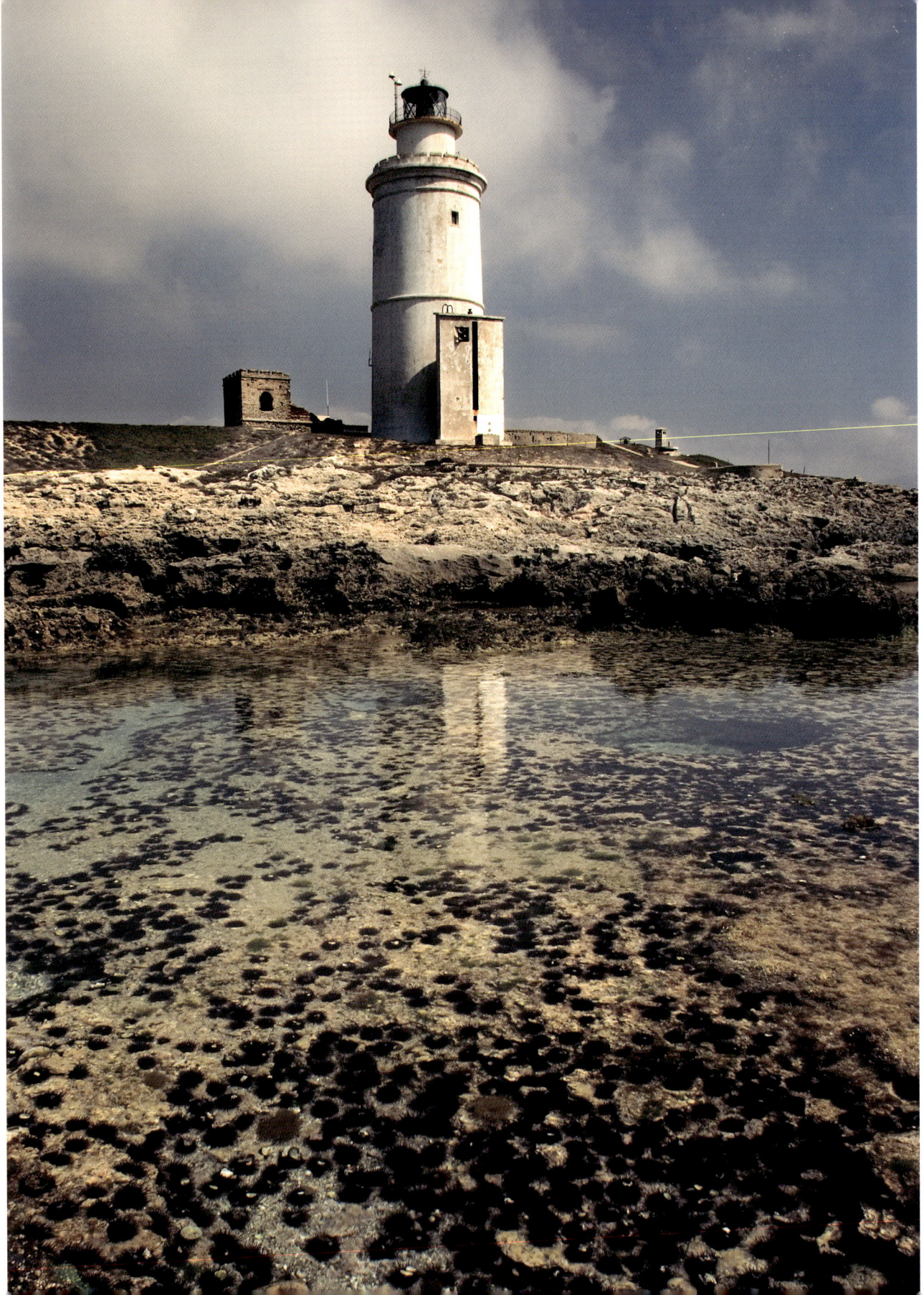

However, the town's principal assets are the surprising variety of habitats within the municipality, from the marshes of the river Palmones to the leafy woods with their mixture of cork oaks, Mirbeck's oaks, alders etc.

As it is very near the city of Algeciras it is fast becoming a commuter town with good services, where people lead a quieter life, not far from their workplace and prefer it to the larger city which is better equipped but where life is more agitated and you get less rest.

San Roque was founded in 1704 by Gibraltarians expelled from Gibraltar after it had been taken over by Anglo-Dutch forces. They situated their new town on a hill, not far from the Rock and near a hermitage dedicated to San Roque. Along with Jimena it is an Andalusian town *par excellence* with its old quarter still intact and it was declared a national monument as a major example of southern Spanish architecture.

For those who play golf or polo this area must be like a dream come true. In nearby Sotogrande, a development with great respect for the environment, polo tournaments are held all year round, with international matches in July and August. It is in fact regarded as the European capital of this sport. As for golf, well, the Ryder Cup and the World Golf Championship were held here at the Valderrama golf course!

San Roque itself is the town in the area with most architectural monuments such as the XVIIth century church of Santa María La Coronada, the hermitage of San Roque, the Governor's Palace, the bull ring, the Town Hall etc.

Not far away is Carteia, founded by the Phoenicians and site of an important Roman settlement; also the Roman villa of Barbésula, hardly excavated at all because of speculative pressure on the land.

The infamous *levante* wind, which throughout the town's history prevented Tarifa from attracting outside investment, especially in the 60's when the boom in tourism began, has, since the 80's, been the economic driving force behind its transformation into a national and international wind surfing centre and is now associated with any sport which requires wind.

The road between Algeciras and Tarifa offers some beautiful scenery with impressive views of the Strait of Gibraltar, but marred on both sides of the road by some hideous-looking power lines, which, to my knowledge, have never been publicly repudiated.

Tarifa is historically famous for the gesture of Guzmán "the Good", governor of the castle in 1295, who refused to give in to the Moors when the castle and its people were besieged, even though it cost the life of his own son, who was in enemy hands. This is the most visited monument in the province and is very well preserved.

The ancient Roman city of Baelo Claudia, beside the stunning beach at Bolonia, was famous for its salted fish industry and for *garum*, a famed sauce which was exported to the Ancient World. The ruins are still being excavated and instead of *garum* you can sample the excellent fish in the area at restaurants which look out onto the beach.

The island of Tarifa in October is an ideal spot to watch migrating seabirds from. Fortunately the entrance to the military zone is restricted, thanks to a wise decision! There you can see what rock pools in the area used to be like until the '70´s − full of life with algae, sea urchins, limpets, snails, crabs, anemones, etc, etc. − in sad contrast to the rest of the coastline where life seems to have deserted these pools. Those of us who knew the coast in the '50´s are

Opposite

Lighthouse on the Isla de las Palomas, Tarifa. In the foreground a rockpool lined with sea urchins.

Punta Paloma beach, Tarifa. In the background on the right Cape Espartel in Morocco.

Female ape with young in Gibraltar.

saddened and deeply concerned about the future of this and other national parks, and Nature in general.

The City of Gibraltar and the Rock of the same name still hold their fortress-like appearance and there is a border crossing between it and the town of La Línea de la Concepción. Once you cross the border you will be surprised at having to cross the airport runway to get to Gibraltar, although this is only the first of many more surprises in store in this city with its wealth of different cultures, religions and ethnic communities.

Another agreeable surprise awaits you when you hear the local inhabitants speaking Spanish with a peculiar Andalusian accent and interspersing words in English: they might say to you "Thank you, *hijo*" or "Bye, Charlie, *nos vemos* tomorrow", without batting an eyelid.

All visitors should go to the museum and the Botanical Garden, as well as wandering down Main Street and other nearby lanes where you will find shops selling spirits, photographic equipment, jewellery and many other things at prices much lower than on the Spanish side of the border.

The Rock is 434 m high and you can reach the top either by cable car or walking up a narrow road where you come to a viewing point with incredible views over the Bay, the Strait and as far as the Costa del Sol. Nearby is the fascinating St Michael's Cave, not forgetting the famous apes and of course it's a unique place from which to watch migration.

Nature, archaeology, gastronomy, fiestas etc.

It's not easy for the visitor to find an area which offers so many possibilities for enjoying subjects related to Nature such as Botany, Zoology and Archaeology. Then there's travel, sport, gastronomy, fiestas and, of course, human relationships.

Organized trips of ornithologists come to the area from, for those of us who live here, countries as far away as Sweden or Finland to see bird migration *in situ*, but I can't understand why other people don't come to see the flowers in bloom. At least I haven't heard of botanists coming here for that sole purpose, as ornithologists do. In April and May you can see major

On the east face of the Rock there is a huge 300 metre high dune.

endemic plants and really amazing flora, both in limestone soils (the Rock and the north-east of Los Alcornocales) and in the sandstone soils (the rest of the area).

In the mid-eighties together with a German friend, who had just arrived in the Jimena de la Frontera area, I was walking through a meadow on limestone soil called Las Cañillas, where the flora was different from the usual Los Alcornocales flora. Suddenly, my friend warned me to watch out as I was about to step on an orchid. I remember I stopped, looked down and asked him, "Orchid? What orchid?"

I was amazed when he pointed out a tiny flower, that I later learned was a *Ophrys tenthrediniphera* and it was then I realized that there was more to the world than hairy, feathery and scaly creatures and it was time for me to stop treading on Nature's marvels because of my ignorance.

The Natural Parks of the Strait, Los Alcornocales and the Upper Rock Nature Reserve in Gibraltar are places where you can see not only major endemic plants but also such little known bird species, even for expert ornithologists, as the White-rumped Swift. The migration of both birds and marine mammals through the Strait of Gibraltar is what gives the area a touch of distinction when compared to other protected areas in Europe.

We still have kilometre upon kilometre of fine, white, sandy beaches with no ugly buildings to spoil them, plus isolated rocky inlets, all in the central area of the windswept coast of the Strait, suitable for almost any sport that requires a sail.

Of archaeological interest are the Arab castles in Tarifa, Castellar de la Frontera and Jimena de la Frontera; then there's the Roman ruins of Baelo Claudia at Bolonia and Barbésula, the Greek colony at Carteya, sacked in 171 B.C. by

The killifish is a small freshwater fish, endemic in the area and whose survival offers some doubt.

Scipio the African, a famous Roman general. And, of course, there are the caves and shelters containing cave paintings, representative of Southern Art.

Last but not least of the attractions of this area we have a varied choice in gastronomy, especially fish freshly caught in the Strait and a plethora of fiestas both national and local, which make visitors think that they've come to a privileged area of the planet, where most of its inhabitants think the rest of the world is the same and don't know how to appreciate what Nature has given them. This modest book is intended to show them.

We have written about all the cultural, industrial, tourist and natural wealth of the Campo de Gibraltar, but to my mind, the least exploited wealth and least considered by the local authorities and what in future, if investment programmes change, could put the area at the

The rhododendron is perhaps the prettiest and most popular plant in the Natural Park of Los Alcornocales.

A fairly large flock of Glossy Ibis flying over the rice fields of La janda.

head of so-called "quality" tourism in Europe and, possibly, in the world, is precisely the observation of the migratory phenomenon. Woods, beaches, monuments, industry, gastronomy, fiestas etc... you've got them in every country, but the migratory spectacle of thousands of soaring birds crossing the skies is restricted to only a few places in the world. Very few indeed, I might add. In Europe it is only comparable to the Bosphorus in Turkey, but here migrating birds disperse over a wide front as it is only one kilometre long and meteorological conditions are not the severe ones we have in Gibraltar. Besides, the birds fly very high over the bustling, sprawling city of Istanbul. An interesting alternative is Eilat in Israel but the uncertain political situation makes many potential visitors not want to run the risk of going there for the migration. For these reasons the Strait of Gibraltar is the most viable alternative for European and North American ornithologists and if we are able to attract them, this will not just be of economic benefit to the area, but also of social and cultural benefit and that is exactly what the Campo de Gibraltar needs. Another thing visitors should bear in mind, and touched on briefly already, is that, thanks to its geographical situation, not only can you watch birds here but also visit conservation areas containing spectacular flora, fauna and countryside, bathe at white, sandy beaches with no built up areas to spoil your enjoyment, and so on and so on.

appendix

The nature photographer's code of practice

1. The welfare of the subject and the conservation of its surroundings are always more important than the photograph.

2. The photographer should be sufficiently familiar with the history and behaviour of the subject, to avoid unnecessary disturbance. He should obtain sufficient technical knowledge so as not to put any species to be photographed at risk in any situation.

3. Always obtain permission from the relevant authorities when the photographing of certain species and their surroundings is required by law. This applies also to the owners of private land. The way of life of people who live and work in the countryside should be respected.

4. To photograph wildlife the photographer should work with species that are wild and free in their natural surroundings without altering their normal behaviour. He should avoid delicate situations such as animals incubating or with newly born young, especially in bad weather (cold, rain, hot sun etc.). If conditions allow photography, maximum precautions should be taken, but if the young are in any danger whatsoever, the attempt should be abandoned.

5. Wherever possible the removal of species for posed studio photography should be avoided. In any case they should be returned to their original habitat unharmed as quickly as possible. This does not include protected species, which depend exclusively on a permit from the relevant authorities.

6. When photographing plants, these should not be partially or totally uprooted and endangered plants should not be touched.

7. The fact that a photograph has been taken under controlled conditions should always be divulged. Photographs in zoos, wildlife centres etc. may well benefit rare and vulnerable wild species by leaving them undisturbed.

8. Avoid cutting branches, twigs and vegetation to camouflage hides used to photograph wildlife from. Use nets made of artificial leaves or, failing that, dead branches and dry vegetation.

9. After manipulating a nest's natural camouflage for photography, this must be restored to its original state. Branches should be tied back rather than cut and the nest, of course, should never be left exposed to predators, people or bad weather.

10. Avoid handling any mineral or archaeological element which might irremediably alter the integrity of a geological or palaeontological formation.

11. Always be discreet when working so as not to attract the attention of the general public or of a predator. Do not reveal the site location of a rare or threatened species, except to accredited scientists or the relevant authorities concerned with its conservation.

12. Always keep your area of work clean and afterwards eliminate all traces of your presence.

13. A nature photographer working abroad should always be as careful and responsible as he would in his own country.

14. Any offence committed against Nature, including illegal actions by other photographers should be reported to the relevant authorities.

15. Please cooperate with your colleagues to improve working conditions in Nature photography and make this code of practice available to anyone who may be unaware of its existence.

The Beaufort Scale

Empirical measure for describing wind intensity, based mainly on the conditions of the sea, its waves and the force of the wind. Its full name is the Beaufort wind force scale.

Beaufort number	(*) Wind Speed	(*) Knots	Description	Sea conditions	Land conditions
0	0-1	< 1	Calm	Flat	Calm. Smoke rises vertically.
1	2-5	1-3	Light air	Ripples without crests.	Wind motion visible in smoke.
2	6-11	4-6	Light breeze	Small wavelets. Crests of glassy appearance, not breaking.	Wind felt on exposed skin. Leaves rustle.
3	12-19	7-10	Gentle breeze	Large wavelets. Crests begin to break; scattered whitecaps	Leaves and smaller twigs in constant motion.
4	20-28	11-16	Moderate breeze	Small waves.	Dust and loose paper raised. Small branches begin to move.
5	29-38	17-21	Fresh breeze	Moderate (1.2 m) longer waves. Some foam and spray.	Smaller trees sway.
6	39-49	22-27	Strong breeze	Large waves with foam crests and some spray.	Large branches in motion. Whistling heard in overhead wires. Umbrella use becomes difficult.
7	50-61	28-33	Near gale	Sea heaps up and foam begins to streak.	Whole trees in motion. Effort needed to walk against the wind.
8	62-74	34-40	Gale	Moderately high waves with breaking crests forming spindrift. Streaks of foam.	Twigs broken from trees. Cars veer on road.
9	75-88	41-47	Strong gale	High waves (2.75 m) with dense foam. Wave crests start to roll over. Considerable spray.	Light structure damage.
10	89-102	48-55	Storm	Very high waves. The sea surface is white and there is considerable tumbling. Visibility is reduced.	Trees uprooted. Considerable structural damage.
11	103-117	56-63	Violent storm	Exceptionally high waves.	Widespread structural damage.
12	118-más	64-71>	Hurricane	Huge waves. Air filled with foam and spray. Sea completely white with driving spray. Visibility greatly reduced.	Considerable and widespread damage to structures.

(*) *Wind speed expressed in km/h and knots in nautical miles/h*

Migratory phenology in the Strait of Gibraltar

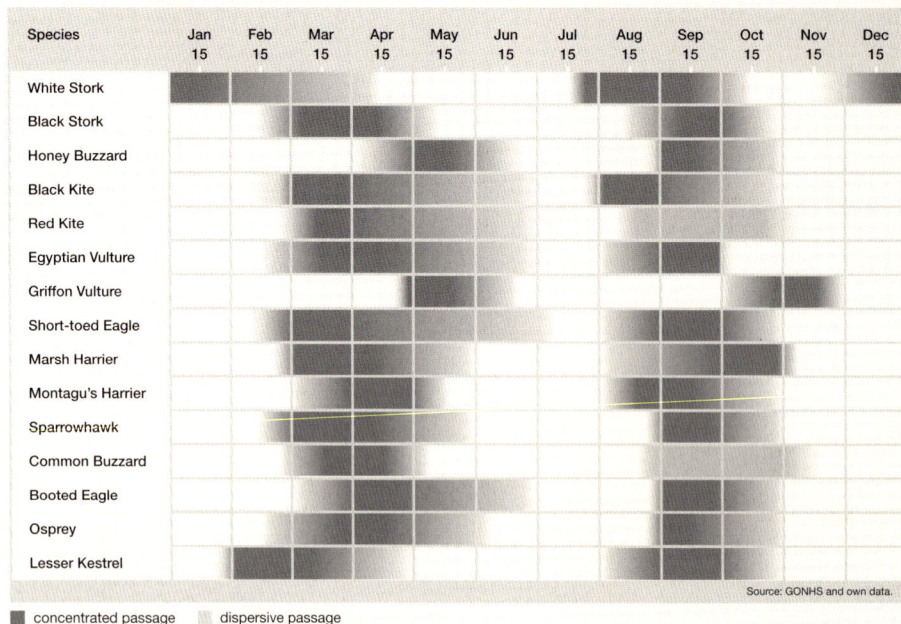

Species	Jan 15	Feb 15	Mar 15	Apr 15	May 15	Jun 15	Jul 15	Aug 15	Sep 15	Oct 15	Nov 15	Dec 15
White Stork												
Black Stork												
Honey Buzzard												
Black Kite												
Red Kite												
Egyptian Vulture												
Griffon Vulture												
Short-toed Eagle												
Marsh Harrier												
Montagu's Harrier												
Sparrowhawk												
Common Buzzard												
Booted Eagle												
Osprey												
Lesser Kestrel												

Source: GONHS and own data.

■ concentrated passage   ▨ dispersive passage

Phenological table of migrants recorded at Hawk Mountain Sanctuary

Species	Average annual count	Record one year count	Record one day count	Peak passage period
Black Vulture / Coragyps atratus	50	80 - 1999	21 - 15/11/1998	November
Turkey Vulture / Cathartes aura	250	367 - 1999	80 - 24/10/1994	late October / November
Osprey / Pandion haliaetus	380	869 - 1990	175 - 23/9/1989	September / early October
Bald Eagle / Haliaetus leucocephalus	60	211 - 2003	48 - 4/9/1948	late August / September
Hen Harrier / Circus cyaneus	230	475 - 1980	36 - 29/10/1955	mid September / mid November
Sharp-shinned Hawk / Accipiter striatus	4350	10.612 - 1977	2475 - 8/10/1979	September / mid November
Cooper's Hawk / Accipiter cooperii	350	1.118 - 1998	204 - 8/10/1981	mid September / mid November
Northern Goshawk / Accipiter gentiles	70	347 - 1972	64 - 10/11/1973	late September/ mid December
Red-Shouldered Hawk / Buteo lineatus	275	468 - 1958	148 - 19/10/1958	October / mid November
Broad-winged Hawk / Buteo platypterus	8230	29.519 - 1978	11.349 - 16/10/1948	12-25 September
Swainson's Hawk / Buteo swaisonii	irregular, 13 total in 11 different years			
Red-tailed Hawk / Buteo jamaicensis	3310	6.208 - 1939	1.144 - 24/10/1939	October / November
Rough-légged Hawk / Buteo lagopus	9	31 - 1961	7 - 11/11/1961	November
Golden Eagle / Aquila chrysaetos	55	159 - 2003	31 - 20/11/2003	late October / November
American Kestrel / Falco sparverius	170	835 - 1989	168 - 3/10/1977	September / early October
Merlin / Falco columbarius	50	176 - 2001	36 - 10/10/1997	late September/ October
Gyrfalcon / Falco rusticolus	irregular, 6 total in 5 different years			
Peregrine Falcon / Falco peregrinus	25	62 - 2002	31 - 5/10/2002	late September / October

Source: Keith L. Bildstein, Sarkis Acopian Director of Conservation Science.

Local ornithological associations

Such a special place in the world with so much biodiversity and ecosystems is bound to be the subject of study and protection both by local people and institutions and by people from other regions. So it's not surprising to find conservationist groups who look after, denounce, study and fight to preserve and sing the praises of these areas of Nature that are under threat because of greed and, what is even worse, ignorance and lack of culture.

GONHS. *Written by Dr. John Cortés*
The Gibraltar Ornithological & Natural History Society, known as GONHS, originated when, at a time when the border with Spain was closed, a group of very young friends went up the "Rock" as Gibraltar's highest part is called, to spend time in its only more or less natural environment. They started bird watching in the 60's and their observations were centred on this very small corner of the Strait, giving them a very deep knowledge of what was happening around them. If we add to this the British ornithological tradition, from Irby to Lathbury, who was one of the most well-loved of Her Majesty's Governors on the Rock, then we have all the ingredients for the creation of what is now a serious and internationally renowned organization.

Its current activities go far beyond mere ornithological observation, although these are still very important. The monitoring both of the migration of soaring birds, of which data exist almost continuously since the early 60's, and of sea birds and also scientific ringing, all continue to form part of the basic aims of GONHS. The data used in Chapter III are the fruit of the work of many ornithologists, both from Gibraltar and abroad, who have worked for this NGO.

There are other important activities in fields such as botany, entomology, marine studies and biodiversity.

This organization also works in a consultancy capacity for the civil and military authorities of Gibraltar in areas concerning the environment and the management of protected areas. It has a representative on several official committees and is employed by the Government to manage the apes and keep the Yellow-legged Gull population under control.

But as a non-profit making society with about 400 members it also organizes trips, monthly talks and other activities that include taking part in events organized by BirdLife International, of which GONHS is a Partner.

GOES. *Written by David Cuenca*
Nearly 23 years ago, a group of nature lovers - especially bird lovers - from the Campo de Gibraltar, led by Alfonso J. González Carbonell, decided to take their hobby a step further and become committed environmentalists, working together to defend nature. Up to then we had been content to go bird watching in the countryside where we might happen to meet up with other colleagues. But now the aim was to preserve and promote our natural surroundings mainly by using birds.

After several meetings at different founders' homes the Grupo Ornitológico del Estrecho (GOES) came into being on 3rd August 1982 in La Línea de la Concepción with a total of just 8 members. It was officially registered (nº 1280) in Cádiz on October 14th that year and is now the oldest ornithological and conservationist association in the Campo de Gibraltar area, playing an important environmental role there and in the surrounding areas.

As it says in our statutes the main aim of the association is to preserve and promote the Environment. Likewise our activities concentrate on the study and protection of wild birds and their habitats, especially vulnerable ones such as wetlands. As the name of our association (Ornithological Group of the Strait) indicates, birds are the tools we use to carry out this work. The ringing of birds is one of our most important activities and this goes back to 1983 when within GOES the Scientific Ringing Team Milvus-Goes was created. Over the last 22 years some 80,000 birds have been ringed and we are one of the most active groups in Spain.

For those of you who don't know what ringing is and why it is useful, let us describe it briefly as the capture and individual marking of a bird so that it can be documented if recaptured. We therefore learn about its movements and are able to trace migratory routes, know where resting sites are, the dates of the different periods of the life cycle of different species (phenology) and their breeding and wintering areas. It is also possible to find out information on population such as rates of survival or how successfully they have bred, check their fidelity to breeding, wintering or passage areas, analyze the difference in migratory behaviour according to sex, age and population, detect changes in populations of a certain species over a period of time, estimate migration speed as well as length of stay in rest areas. Finally, ringing birds helps us know their biometry, body health and causes of death.

Over the years many people have joined and left the group, a few have stayed and even fewer are original founding members. Those who left perhaps didn't find what they expected to find. Ideological differences have been and still are causes for disagreements, as in any collective and we are no exception although we try to improve in spite of them. Nevertheless back in 1986 GOES almost disappeared because of the differing currents of opinion between the "radical ecologists" and the "ringing" group. There was a split, the former went their own way but the ringers carried on thanks to a few stalwarts, one of them being Javier Espinosa, now a veteran of the association and its current president. He decided at the time to gather up all the association's material and take it home for safekeeping.

We have had many successes over the years but there is still a lot to be done. Several studies about birds in the Campo de Gibraltar have been published such as Gyps, Informes Ornitológicos del Estrecho and the Milvus of which we are preparing the sixth edition with a summary of ringing work carried out between 1994 and 2003. On many occasions we have given talks in schools, as well as organized regular introductory courses on scientific ringing and bird watching. With the estimable help of the Nature Protection Unit of the Civil Guard we have succeeded in controlling illegal activities related to birds and the environment. More specifically we have reported several illegal encroachments on open spaces to the authorities, thus avoiding the destruction of these valuable areas of the environment. We are also proud of the fact that in 1989 we made a decisive contribution to the inclusion of the three most important wetlands in the Campo de Gibraltar in the Network of Protected Nature Land in Andalusia. The Playa de los Lances in Tarifa, the Estuario del Río Guadiaro in San Roque and the Marismas del río Palmones in Algeciras are all now Nature Areas, thanks to reports made by GOES with its own data and experience.

We have been or are currently involved in the following projects: ringing of White Stork chicks in the province of Cádiz, monitoring of bird populations in the Natural Park of the Sierra de Grazalema and Sierra de las Nieves, the Calidris programme (ringing of sea and shore birds), the Euring programme (ringing of the Swallow), the Passer programme (monitoring common species by ringing), the tarsus programme (evaluation of most suitable ring types) and the monitoring of three types of Wheatear in the sierras of Ronda. We have been ringing birds for over 20 years in several areas such as Palmones and Guadiaro – especially during postnuptial migration and wintering times – and also on large estates such as Guadacorte and La Almoraima. We also work in collaboration with similar-minded associations, although perhaps not as often as we should, like COCN (Black Stork Ornithological Group), Verdemar, Agaden, the Cádiz Natural History Society (SGHN) and GONHS. Also with local government departments in the city councils of La Línea and Algeciras and the provincial delegation of the Dept. of the Environment of the Regional Government of Andalusia, and with companies such as the Port Authority of Algeciras (APBA), Ornitour S.L. and the Migres Foundation which was set up quite recently.

Now, after a few not-so-good years, we are well established. We have about thirty members and many volunteers. Most are from the Campo de Gibraltar, mainly La Línea and Algeciras, but also from the Sierra de Cádiz, Serranía de Ronda and the Costa del Sol. There are also three English ringers who have come to live in the area. Among our members there are 14 ringers, one of whom, Manuel Lobón, belongs to the Ringers Commission, an organism that is part of the Bird Migration Centre which organizes ringing on a national scale. So, things are looking up!

Let me say in conclusion that nature observation can, sometimes, be unrewarding. But only sometimes, so don't give up too easily. You have to enjoy it and you need a lot of patience in all senses: birds and ringing provide unforgettable experiences from seeing a Kingfisher or Golden Oriole for the first time or the passage over the Strait of thousands of birds in a single day, to the capture of a rare species such as the Rustic Bunting – two recently ringed birds were only the third and fourth to be ringed on Spanish soil - , the recovery a Blackcap or a Bluethroat from Belgium, and then the news that a Yellow Wagtail ringed in Palmones has been recovered in Senegal or finding a Blue Tit that was ringed as a chick seven years before (quite a feat for such a small bird).... All this is quite addictive, I can tell you!

COCN. *Official description*

The Black Stork Ornithological Association (COCN) is a voluntary non-profit organization which was incorporated in the Official Association Register of the province of Cádiz on 27th September 1997.

Although that date is when the association became official its members had already been active for some time, either individually or in small groups, involved in the study and conservation of birds in the Campo de Gibraltar area. In fact in the data base of the Ornithological Station in Tarifa there is a lot of information provided by our members from field work and studies of the migration of soaring birds in the Strait of Gibraltar dating back as far as 1983.

The association's main aims are the study, conservation and promotion of nature in general, and birds in particular. These aims are carried out in the form of different projects undertaken voluntarily by members and volunteers.

A major project of ours is the Ornithological Station at Tarifa, a centre open to the public near the Strait of Gibraltar which coordinates many of the group's projects and those of other associations and institutions.

The importance of the Strait of Gibraltar and the projects we are involved with in this area is that, because of diverse geographical, physical and biological circumstances, a large number of birds have to migrate through here. Thus, over a very small surface area we can count a huge number of birds moving between Europe and Africa.

While studying and analyzing this movement of birds and their behaviour in the area we are trying to determine their population numbers as it is vital to know whether numbers are increasing or diminishing. We get to know the environmental state of health of their habitats both in Europe and in Africa. Birds are a useful indicator when we wish to study and conserve biodiversity.

At the moment several ornithological studies are underway in the area, many of them first-time ones and they will be used as bases for future studies. A data base is available to the public and is being used by many researchers. Courses, conferences, exhibitions etc. help to train volunteers and we are trying to make the local population aware of this important resource on their doorstep.

The association is also carrying out several studies of endangered species and the tremendous amount of information being generated by COCN, both on a local, national and international level about the migratory movement of birds in the Strait of Gibraltar and its importance, is attracting more and more volunteers, scientists and ornithologists from all over the world. This is why COCN, realizing the need for them, has set up a Telephone Service of Ornithological Information, opened the Tarifa Ornithological Station etc. etc.

Although all the association's activities are carried out by volunteers, this in no way diminishes either the reliability or the quality of their work and results. Up to now 95% of volunteers involved in COCN field work have either been professionals or students of courses related to the study and conservation of nature. For example, these professionals have provided ideas and solutions for the creation of data bases, planning and design of bird observatories in the Strait of Gibraltar.

The association has grown a lot in just a few years as far as membership, collaborators and volunteers are concerned. But we also get more support now from the public and from the private sector to help in our conservation work, studies and promotion of nature.

MIGRES Foundation. *Official description*

It was set up by the Regional Government of Andalucía towards the end of 2003 with the aim of making economic use of the migratory phenomenon in the Strait of Gibraltar and organizing events, activities and projects related to sustainable development. Since its inception it has always believed that nature conservation should not cost society money, but rather it should represent an opportunity to create wealth and generate resources.

The foundation offers itself as a meeting point for experts, institutions and associations that share a common desire to promote the study, conservation and divulgence of Nature and, especially, the enormous cultural and ecological values present in the Campo de Gibraltar. Furthermore, a large part of its efforts are directed at consolidating and strengthening the MigreS Programme and developing activities of socio-economic relevance that make sustainable use of the natural resources of the area.

The foundation is actively working towards achieving these goals by fostering projects in environmental education (producing plays, publishing books, supporting initiatives etc.), in sustainability and development (such as the sixteen projects of compensatory measures directly related to the second electricity cable link-up between Spain and Morocco, or the environmental projects undertaken with the wind farms in the Wind Energy Association of Tarifa, the pioneering work carried out in land custody – creating, together with another four Andalusian foundations, Insulas : The Andalusian Network of Land Custody and Management - , and socio-economic projects related to natural resources, especially nature tourism) and, of course, the environment and nature conservation which is basically what the MigreS programme is all about.

So as to be certain of maintaining scientific rigour, both in the design and in the evaluation of projects and programmes, the foundation has a committee of internationally renowned scientists and experts in the migratory phenomenon, and whose members are Dr Keith L. Bildstein (Scientific director at the Hawk Mountain Sanctuary, USA.), Prof. Ian Newton (Researcher at the Centre for Ecology and Hydrology, United Kingdom), Dr Lucas Jenni (Scientific director of the Swiss Ornithological Institute, Switzerland) and Dr Miguel Ferrer (Researcher at the Biological Station at Doñana, Spain).

As it is the project that is most closely related to the subject of this book, let's take a brief look at the MigreS Programme.

In 1995 the CMA (Department for the Environment) of the Andalusian Regional Government commissioned the Biological Station at Doñana to define the strategy required to make economic use of migration in the Strait of Gibraltar.

As a result, in 1996 the idea of the MigreS Programme began to take shape, a project aimed at studying the migration of different groups of animals passing through the Strait of Gibraltar and obtaining information about the evolution of these species' populations. The main premise of the programme right from the start was for it to be a long-term one, something which is fundamental when monitoring animal populations and its mainstay is the environmental volunteer, an essential ingredient for the programme to to be able to continue indefinitely.

The different projects of the MigreS Programme all have two separate phases: phase one is to decide on the monitoring methodology and how to optimize efforts and phase two concentrates on monitoring itself.

Thus, in 1997 the CMA commissioned the Spanish Ornithological Society (SEO/BirdLife) with the design of monitoring methodology in the postnuptial migration of soaring birds over a period of 5 years (1997 – 2001). From 2002 to 2005 the same institutions developed a second phase, namely the monitoring itself. In 2002 the University of Cádiz, again commissioned by the CMA, extended the monitoring to include seabirds and shorebirds and the design phase was finished in spring 2004.

In 2005 the foundation took over the coordination of the monitoring phase of seabirds and in 2006 it took over from SEO to coordinate the monitoring of soaring birds. In addition, during that same year it was asked to design the monitoring of passerine migration (mainly by using scientific ringing techniques) and the spring migration of soaring birds.

The foundation now intends to create a Migration Monitoring Station in the Strait of Gibraltar to be used as a base at which to unify and optimize efforts in the development of projects which, by their very nature, are carried out in direct contact with the migratory phenomenon These projects will become part of the MigreS Programme and will include projects currently underway, those that are now in the design phase and those that in the future will deal with new groups of species and possible extensions of activities of all projects (marking, radio tracking, etc). This will allow all projects to be carried out using uniform criteria and to complement each other efficiently, thus achieving optimization of human and material resources.

At the same time the foundation is working actively, looking for financial support from public and private entities, to consolidate and strengthen all these projects.

Acknowledgements

My friend Mariano Vargas, a versatile artist and illustrious specialist in such a difficult subject as nude photography, unintentionally provoked me into writing this book.

My wife Palma del Valle has stood bravely in the firing line while suffering all the ups and downs of the lengthy creation, financing and production process. She never ceased to support me and many of her suggestions have led to notable improvements in the final version.

Manuel Jesús Cabello, the first and current director of the Natural Park of the Strait, devoted his valuable free time to obtaining financial support from the Junta de Andalucía and other organisations to make the publishing of this book a reality. He kept faith with the project right from our very first conversation. Without his estimable help it would have been extremely difficult for me to find that first, essential, economic support.

My Gibraltarian friend, Dr John Cortés, kindly agreed to write Chapter III: The Rock of Gibraltar: another view of migration. An extremely busy man, he took time off from his multifarious activities in order to write and also offer advice on the English translation. Besides, he showed me the most interesting places on the Rock from which to observe migration, as well as, surprisingly, the secrets of nature on limestone in a relatively small space.

Enrique Aguirre and Federico Fuentes started the layout but later had to give up due to overwork and lack of time.

Michael Potts was not only in charge of the English translation but also an expert at spotting errors and a budding ornithologist. Thanks to him, some of the rivers in the Campo de Gibraltar run into the right seas and post- and prenuptial migrations appear correctly in the text.

I am grateful to Dr Keith L. Bildstein for writing the prologue and telling us about Hawk Mountain Sanctuary, where he is currently the director. His vision and management of this very important enclave in North America may serve as a model for us to dignify what we have here.

My brother Manuel Barcell de Arizón contributed a very personal, sensitive text about fungii, which we all love to eat and are a great excuse for getting together and going out into the countryside. His unswerving efforts have opened the doors to several sponsors.

Carmen Bosh, Mariano Vargas and Alfonso del Valle worked on maps, which unfortunately, due to questions of design, have not been included in the book.

Jesús Cabello, Manuel Barcell, Ildefonso Sena, Rafael Peña, Eduardo Briones, Juan Pérez, Martín Caballero and Diego Sánchez Rull put me in touch with institutions and companies that have financed this book.

Behind the company logos and local government coats of arms there are people who supported and believed in the project and I should therefore like to thank Francisco Oñate, Juan Montes de Oca, Fernando Molina, Paola Moreno, Fátima Andrade, Carlos de la Rosa, Gerardo Landaluce, Jesús Maraver, José Botía, Álvaro Rodríguez, Emilio Aragón, Rosa Vázquez, Fernando Alpresa, Federico Fernández, José Luis Romero, Francisco de Paula and Ángel Gavino.

My Galician namesake Fernando Bandín Monteiro, his wife Eva, Michael Potts and Carlos M. García, conscientiously correct proofreaders, have checked the texts and criticized them mercilessly while solving any lapses.

This book is an excellent opportunity for those photographer friends of mine who work in the Strait to be able to show some of their work, so I'd like to thank all of them for having lent me some of the magnificent photographs that dignify the book. They are: Andrés Domínguez, Ángel Sánchez, Antonio Benítez Barrios, Antonio Gavira, Charles Pérez, Christopher Buttigieg, Enrique Aguirre, Fernando García Arévalo, Guillermo Doval, José Luis Roca, Leslie Linares, Manolo Castro, Manolo Rojas, Marcos G. Meider, Paco Vega, Palma del Valle, Paul Acolina, Teo Todorov and, last but not least, Fernando Barrio Fuentenebro, who supplied me with the photograph of Francisco Bernis, and Cristina Bernis who authorized its publication.

Thanks also to Manuel Fernández Cruz who, over seven years, let me collaborate with the different teams he directed in their field studies of the migration of the White Stork and to whom I owe most of the knowledge I gleaned about it and express in this book.

José Solera, a meteorologist, checked everything related to his subject on the Campo de Gibraltar.

Lalo Ventoso and Maties Bebassa supplied me with information about migration in the Balearic Isles.

Ana Carnicer and Luís Maraia, designers and layout artists of the book, have made my work easier at all times and their artistic contribution is very important indeed.

Gaspar, David, Andrea, Johanna, Maribel and the rest of the personnel at Grafisur printers in Tarifa have collaborated enthusiastically so that the book is of the best possible quality.

Enrique Navarro agreed to take part in the project with his magnificent drawings and did these in record time. This last-minute participation has added an extra dimension to this work.

In one way or another different friends and relatives have also collaborated and I am grateful for their help both as regards comments they have made and as companions in recent or past excursions into the countryside: Alfonso de la Cruz, Ana Belén Barrios, Ana Juárez, Andrés Domínguez, Ángel M. Sánchez, Ana María Gutierrez Serrano, Ana Retamero, Antonio Aguilar, Antonio Sabater, Asunción Domínguez, Blanca Román, Bob Wheeler, Carlos M. García, Carlos Riera, Charles Pérez, Clive Finlayson, Conchita Barrios, Cristina Parkes, Cristóbal García, Dani Martínez, Dany Blanco, David Barros, David Cuenca, David Rios, Eduardo Briones, Emilio Parejo, Enrique Aguirre, Enrique Salvo, Fernando Bandín, Fernando Ortega, Francisco Montoya, Francisco Solera, Harry Vangils, Iñigo Sánchez, Jaime Nieto, Javier Espinosa, Javier Navarro, Jesús Parody , Joaquín Gutiérrez Acha, Joaquín Mazón, Juan Ramírez, John Cortés, José Luis Márquez, José Luis Paz, José María de Francisco, José María Lubián, José Ramón

Benítez y Marisa, Keith Bensusan, Leslie Linares, Lola Cabrera, Luis Valverde, Manolo Castro, Manuel Barcell, Manuel Español, Manuel Fernández Cruz y Paz, Manuel Lobón, Marcos G. Meider, Martín Caballero, Michael Potts, Miguel Ángel Quevedo, Pablo Ortega, Paco Montoya, Paco Solera , Palma del Valle, Paul Acolina, Paul Roca, Pedro Montoya, Pepe Álvarez, Pilar del Cañizo, Rosa Molina Gil, Vicente Asensio, Sebastián Zarza, Teo Todorov.

Index of guest photographers

Bibliography

Barrios, F. (1993). *Vencejo Cafre. Vivir en casa ajena.* La Garcilla 87: 22-23. Madrid.

Barrios, F. (1994). *Primeros datos sobre reproducción del Vencejo Cafre en España.* Quercus 95: 6-8.

Barrios, F. *Vencejo Cafre. Llegadas y ocupaciones.*

Barrios, F. (1994). *Gibraltar. Asociación Española de Fotógrafos de Naturaleza.* IRIS.

Barros, D. y Ríos, D. (2002).*Guía de las aves del Estrecho de Gibraltar.*

Bernis, F. (1959). *La migración de las cigüeñas españolas y las otras cigüeñas "occidentales".* Ardeola 5: 93-114.

Bernis, F. (1971). *Aves migradoras ibéricas.* 8 fascículos. Sociedad Española de Ornitología. Madrid.

Bernis, F. (1973). *Migración de Falconiformes y Ciconia spp. por Gibraltar.* Verano-otoño 1972-1973. Primera parte. Ardeola 19: 151-224.

Bernis, F. (1980). *La migración de las aves en el estrecho de Gibraltar.* Vol. I Aves planeadoras. Universidad Complutense de Madrid.

Bernis, F. (1981). *La población de las cigüeñas españolas.* Universidad Complutense. Madrid.

Bernis, F. (1988). *Los vencejos. Su biología, su presencia en las mesetas españolas como aves urbanas.* Universidad Complutense. Madrid.

Bernis, F. y García Rúa, A. (1974). *Observación Buteo rufinus en la provincia de Cádiz.* Ardeola 20: 341-343.

Bildstein, K. (2006). *Migrating Raptors of the World Their Ecology and Conservation.*

Ceballos, J. y Guimerá, V. (1992). *Guía de las aves de Jerez y de la provincia de Cádiz.* Biblioteca de Urbanismo y Cultura, Jerez.

Cortés, J. E.; Finlayson, J. C.; Mosquera, M. A. ; & García, E. F. J : (1980). *The Birds os Gibraltar.* Publisher by the Gibraltar Bookshop. Gibraltar.

Del Hoyo, J.; Elliott, A. & Sargatal, J. eds (1994). *Handbook of the Birds of the Wordl.* Vol. II. New World Vultures to Guineafowl. Lynx Eds. Barcelona.

Del Junco, O. y Barcell, M. *El buitre leonado (Gyps fulvus) en Cádiz.* Junta de Andalucía. Consejería de Medio Ambiente.

Elphick, J. (1995). *Aves. Las grandes migraciones.* Encuentro Editorial, S. A.

Étienne, P y Carruette, P. (2004). *La cigüeña blanca.* Ediciones Omega, S. A.

Finlayson, J. C. (1992). *Birds of the Strait of Gibraltar.* T & A. D. Poyser. London.

Ford, R. (1980). *Manual para viajeros por Andalucía y lectores en casa.* Ediciones Turner.

Gómez, F. y Díaz, J. L. (1991). *Guía de los peces continentales de la Península Ibérica.* Ed. Acción Divulgativa.

Gutiérrez, J.M.; Martín, A.; Domínguez, S.; Moral, J.P. (1991). I*ntroducción a la Geología de la Provincia de Cádiz.* Universidad de Cádiz.

Hilgerloh, G. (1988). *Radar observations of passerine trans-Saharan migrants in southern Portugal.* Ardeola 35 : 41-51.

Lázaro, E. & Fernández Cruz, M. 1989. *Migra-*

tion of the White Store through the strait of Gibraltar in 1985: first results. En, G. Rheinwald, J. Orden & H. Schulz (Eds.): Weisstorch: Status and Schutz, pp. 263-268. Proc. 1st Int. Stork Conserv. Symp., Walsrode, 1985, Schriftenr., Braunschweig (Alemania).

Martí, R. & Del Moral, J. C. (Eds) (2003). *Atlas de las aves reproductoras de España.* Dirección General de Conservación de la Naturaleza-Sociedad Española de Ornitología.

Molina, B. & Del Moral, J. C. 2005. *La Cigüeña Blanca en España.* VI Censo Internacional (2004). SEO/BirdLife. Madrid.

Parejo, E. L. y Sáez, O. (1995). *Estudio ornitológico del Campo de Gibraltar y Ceuta.* Instituto de Estudios Campogibraltareños. Cádiz.

Pérez-Petinto, M. (2004). *Historia de Algeciras.* Instituto de Estudios Campogibraltareños. Mancomunidad de Municipios del Campo de Gibraltar.

Porter, R. F.; Willis, I.; Christensen, S. & Nielsen, B. P. (1981). *Flight Identification of European Raptors.* T. & A. D. Poyser. Calton.

Rodríguez, A.; Delibes, M. y Palomares, F. (2003). *Lince Ibérico Bases para su reintroducción en las sierras de Cádiz.* Consejería de Obras Públicas y Transportes Gestión de Infraestructuras de Andalucía, S. A. (GIASA).

SEO (1998). *Programa Migres. Seguimiento de la Migración en el Estrecho.* Proyecto Piloto 1997. Consejería de Medio Ambiente de la Junta de Andalucía y SEO/BirdLife. Madrid.

SEO/BirdLife (1999). *Programa Migres. Seguimiento de la Migración en el Estrecho.* Consejería de Medio Ambiente de la Junta de Andalucía y SEO/BirdLife. Madrid.

Tellería, J. L. (1981). *La migración de las aves por el Estrecho de Gibraltar. Vol. II Aves no planeadoras*: Universidad Complutense de Madrid. Madrid.

Tellería, J. L.; Díaz, M. y Asensio, B. (1996). *Aves Ibéricas. Vol. I, no paseriformes.* J. M. Reyero Editor. Madrid.

Tellería, J. L.; Díaz, M. y Asensio, B. (1999). *Aves Ibéricas. Vol. II, paseriformes.* J. M. Reyero Editor. Madrid.

Varios (1994). *III Jornadas de Historia del Campo de Gibraltar.* L Línea de la Concepción. Mancomunidad de Municipios del Campo de Gibraltar.

Varios (1996). *II Jornadas de Estudio y Conservación de la Flora y Fauna del Campo de Gibraltar.* Jimena de la Frontera. Almoraima. Mancomunidad de Municipios del Campo de Gibraltar.

Varios (1998). *II Jornadas de Estudio y Conservación de la Flora y Fauna del Campo de Gibraltar.* Castellar de la Frontera. Almoraima. Mancomunidad de Municipios del Campo de Gibraltar.

Epilogue

Over the course of eight chapters we have accompanied readers across old Europe, the vicissitudes of the Strait of Gibraltar, the arid Sahel and even as far as Hawk Mountain Sanctuary in the USA ....along innumerable, imaginary routes that become blurred with the passage of Nomads of the Strait of Gibraltar, and also through the conservation areas of Gibraltar, the Natural Parks of Los Alcornocales and the Strait that have revealed to both the casual visitor and to experts such stunning, singular biodiversity.

I will be happy, indeed very satisfied, if your reading of this book has achieved the goal I set, namely to introduce you to a privileged place within the western Palaearctic: the European coast of the Strait of Gibraltar and nearby enclaves stretching from Fuente de Piedra in Málaga to the internationally famous National Park of Doñana.

This is the first book to narrate and illustrate in detail the migratory phenomenon along the European coast of the Strait. It is something which the public in general is not aware of, but can and should be seen as an opportunity to attract ornithologists, naturalists and researchers from other parts of the country and abroad. In 1972 Francisco Bernis began studying migration in the Strait, yet now, 34 years later we are still intrigued by many aspects of migrants' behaviour – there lies a golden opportunity for those researchers and aficionados to try and explain them!

Every year more and more European ornithologists come down to watch migration, but there are still many who are unaware of the real magnitude of this phenomenon and they must be informed and then convinced to visit the Strait. This is the responsibility of the local and regional administrations in Andalusia. The Regional Government of Andalusia took the first step some years ago and now it is time for town councils, municipal associations and businessmen to continue the work at trade fairs and international exhibitions. The reasoning is simple: there are local fairs, sightseeing, beaches, cities etc, all over the planet, but places to witness the spectacular migration of large soaring birds can be counted on the fingers of one hand – and one of them is here.

Printed and bound
by Grafisur,
in the city of Tarifa,
March 2007.